The Chapin-Stephens Co. was created in 1901 through a merger of H. Chapin's Son Co. and D.H. Stephens & Co. The Chapin company began as a partnership between Hermon Chapin and Daniel Copeland, formed in 1826 at Pine Meadow, Connecticut. It was one of the earliest, and ultimately among the largest, 19th century manufacturers of wooden planes, rules, and other hand tools. Delos H. Stephens learned rule-making working for Chapin before striking off on his own in 1854.

With the Chapin interests in control the newly merged company continued as before. As this 1914 catalog illustrates, most products remained unchanged and little new was added. The firm reorganized in 1927 and finally dissolved in 1929, with the Stanley Rule & Level Co. acquiring and absorbing the line of rules but abandoning the manufacture of wooden planes.

CATALOG NO. 114

THE CHAPIN-STEPHENS CO.
UNION FACTORY

FACTORIES AND GENERAL OFFICES
PINE MEADOW, CONN., U. S. A.

NEW YORK OFFICE
126 Chambers Street

THE ASTRAGAL PRESS

Box 338
Morristown, New Jersey 07963-0338

Library of Congress Catalog Card Number 91-70726
International Standard Book Number: 0-9618088-8-8

INDEX.

F. M. CHAPIN, *President.*

H. M. CHAPIN, *Vice-President and Treasurer.*

F. L. STEPHENS, *Secretary.*

Since our consolidation in 1901 of the plants of H. Chapin's Son Company, established in 1826, and Stephens & Company, established in 1854, we have reorganized nearly every department in our factory, moving some of them into new and larger quarters, so that now the combined capacity of our works has been more than doubled.

In making these betterments we have not for a moment lost sight of quality and finish, but rather when it has been possible for us to do so, we have improved same.

It is our purpose to continue to guarantee all goods of our manufacture as has been our practice since this business was established.

To our many customers throughout the civilized world we extend our thanks for the continued preference they have shown for our tools during the nearly ninety years of our existence, as is evidenced by the growing demand for same. We hope to continue to merit this appreciation shown us by the public.

THE CHAPIN-STEPHENS COMPANY,

PINE MEADOW, CONN.

BOXWOOD RULES.

POCKET.

One Foot.   **Four Fold.**

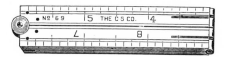

No. 69.   Two-Thirds Size.

No.   Per Dozen
69.   Round Joint, Middle Plates, 8ths and 16ths of inches, ⅝ in. wide $1.75

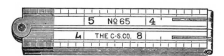

No. 65.   Two-Thirds Size.

65.   Square Joint, Middle Plates, 8ths and 16ths of inches, ⅝ in. wide   2.00

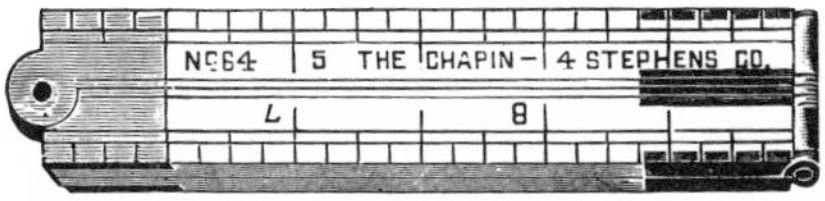

No. 64.   Two-Thirds Size.

64.	Square Joint, Edge Plates, 8ths and 16ths of inches,	⅝ in. wide		2.75
64¼.	Square Joint, Edge Plates, 8ths and 16ths of inches,	⅞ "	"	3.25
65½.	Square Joint, Bound, 8ths and 16ths of inches, .	⅝ "	"	5.50
55.	Arch Joint, Middle Plates, 8ths and 16ths of inches,	⅝ "	"	2.50
56.	Arch Joint, Edge Plates, 8ths and 16ths of inches,	⅝ "	"	3.50

No. 57.   Two-Thirds Size.

57.	Arch Joint, Bound, 8ths and 16ths of inches,	⅝ in. wide		6.25
57½.	Arch Joint, Bound, 8ths and 16ths of inches,	⅞ "	"	7.00

BOXWOOD RULES.

NARROW.

Two Feet.   Four Fold.

No. 68. Two-Thirds Size.

No.   Per Dozen
68.   Round Joint, Middle Plates, 8ths and 16ths of inches, 1 in. wide $2.50

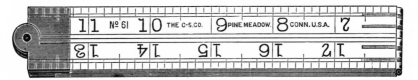

No. 61. Two-Thirds Size.

61.   Square Joint, Middle Plates, 8ths and 16ths of inches, 1 in. wide   3.00

No. 63. Two-Thirds Size.

63.   Square Joint, Edge Plates, 8ths, 10ths, 12ths, and
            16ths of inches, Drafting Scales, ... 1 in. wide   4.00
84.   Square Joint, Half Bound, 8ths, 10ths, 12ths, and
            16ths of inches, Drafting Scales, ... 1 "   "   6.50

BOXWOOD RULES.

NARROW.

Two Feet.   **Four Fold.**

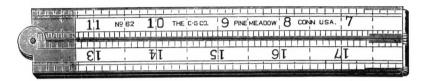

No. 62. Two-Thirds Size.

No.   Per Dozen

62.   Square Joint, Bound, 8ths, 10ths, 12ths, and 16ths of
inches, Drafting Scales, ... 1 in. wide $8.00

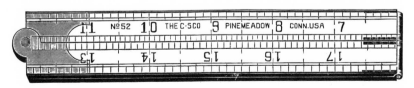

No. 51. Two-Thirds Size.

51.   Arch Joint, Middle Plates, 8ths, 10ths, 12ths, and
16ths of inches, Drafting Scales, ...   1 in. wide   3.50
53.   Arch Joint, Edge Plates, 8ths, 10ths, 12ths, and 16ths
of inches, Drafting Scales, ... 1 "   "   4.50

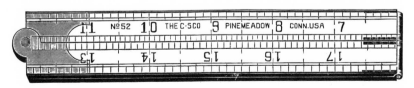

No. 52. Two-Thirds Size.

52.   Arch Joint, Half Bound, 8ths, 10ths, 12ths, and 16ths
of inches, Drafting Scales, ...   1 in. wide   7.25

BOXWOOD RULES.

NARROW.

Two Feet.   Four Fold.

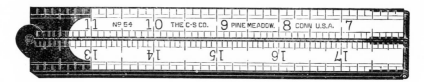

No. 54.   Two-Thirds Size.

No.   Per Dozen

54.   Arch Joint, Bound, 8ths, 10ths, 12ths, and 16ths
        of inches, Drafting Scales, ... 1 in. wide $8.75

54S. Arch Joint, Bound, 10ths and 100ths of a foot, 10ths
        and 16ths of inches, ... 1 "   "   10.00

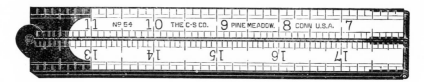

No. 59.   Two-Thirds Size.

59.   Double Arch Joint, 8ths, 10ths, 12ths, and 16ths of
        inches, Drafting Scales, ... 1 in. wide   5.25

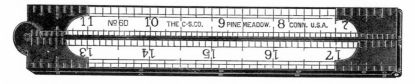

No. 60.   Two-Thirds Size.

60.   Double Arch Joint, Bound, 8ths, 10ths, 12ths, and
        16ths of inches, Drafting Scales, ... 1 in. wide 10.75

BOXWOOD RULES.

BROAD.

Two Feet.　　　Four Fold.

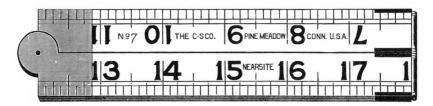

No. 67.　Two-Thirds Size.

No.　　　　　　　　　　　　　　　　　　　　　　　　Per Dozen

67.　Round Joint, Middle Plates, 8ths and 16ths of inches, 1⅜ in. wide　$3.50

No. 70.　Two-Thirds Size.

70.　Square Joint, Middle Plates, 8ths and 16ths of
　　　inches, Drafting Scales,　. . . . . .　1⅜ in. wide　4.00
72.　Square Joint, Edge Plates, 8ths, 10ths and 16ths
　　　of inches, Drafting Scales,　. . . . .　1⅜ "　"　5.00

No. 7.　"Nearsite."　Two-Thirds Size.

7.　Square Joint, Edge Plates, 8ths and 16ths of inches,
　　　with Large Red Figures, ...　. .　1⅜ in. wide　11.00
72¼.　Square Joint, Half Bound, 8ths, 10ths and 16ths
　　　of inches, Drafting Scales,　. . . . .　1⅜ "　"　8.00
72½.　Square Joint, Bound, 8ths, 10ths and 16ths of
　　　inches, Drafting Scales, ...　1⅜ "　"　9.00
73.　Arch Joint, Middle Plates, 8ths, 10ths and 16ths
　　　of inches, Drafting Scales, ...　1⅜ "　"　5.00

BOXWOOD RULES.

BROAD.

Two Feet.   Four Fold.

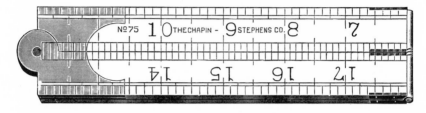

No. 75.   Two-Thirds Size.

No.   Per   Dozen

75. Arch Joint, Edge Plates, 8ths, 10ths and 16ths of
        inches, Drafting Scales, ... 1⅜ in. wide $6.00

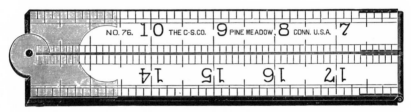

No. 76.   Two-Thirds Size.

76. Arch Joint, Bound, 8ths, 10ths, and 16ths of inches,
        Drafting Scales, ... 1⅜ in. wide   10.00
76½. Arch Joint, Half Bound, 8ths, 10ths, and 16ths of
        inches, Drafting Scales, ... 1⅜ "   "   8.50
77.   Double Arch Joint, 8ths, 10ths, and 16ths of inches,
        Drafting Scales, ... 1⅜ "   "   6.50
78.   Double Arch Joint, Half Bound, 8ths, 10ths, and
        16ths of inches, Drafting Scales, ... 1⅜ "   "   10.00

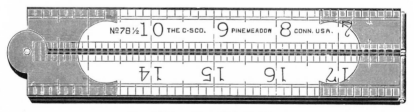

No. 78½.   Two-Thirds Size.

78½. Double Arch Joint, Bound, 8ths, 10ths, and 16ths
        of inches, Drafting Scales, ... 1⅜ in. wide   12.00
83.   Arch Joint, Edge Plate, *Slide,* 8ths, 10ths, and 16ths
        of inches, Drafting Scales, ... 1⅜ "   "   10.00

BOXWOOD RULES.

BOARD MEASURE.

Two Feet.   **Four Fold.**

No.   Per Dozen

79.   Square Joint, Edge Plates, 12ths and 16ths of
       inches, Drafting Scales, ... $1\frac{3}{8}$ in. wide $7.00
79¼. Square Joint, Bound, 12ths and 16ths of inches,
       Drafting Scales, ... $1\frac{3}{8}$ "   "   11.00
81.   Arch Joint, Edge Plates, 12ths and 16ths or inches,
       Drafting Scales, ... $1\frac{3}{8}$ "   "   8.00
82.   Arch Joint, Bound, 12ths and 16ths of inches,
       Drafting Scales, ... $1\frac{3}{8}$ "   "   12.00

No. 79½.   Two-Thirds Size.

79½. Square Joint, Edge Plates, with Board Stick Table,
       8 lines, 12 to 22 feet, ... $1\frac{3}{8}$ in. wide   7.00
79¾. Square Joint, Edge Plates, with Board Stick Table,
       8 lines, 12 to 22 feet, T Head, ... $1\frac{3}{8}$ "   "   11.00

Two Feet.   **Two Fold.**

22.   Square Joint, 10ths, 12ths, and 16ths of inches,
       Octagonal Scales, ... $1\frac{1}{2}$ "   "   5.00
23.   Arch Joint, 12ths and 16ths of inches, Octagonal
       Scales, ... $1\frac{1}{2}$ "   "   7.00
24.   Arch Joint, Bound, 12ths and 16ths of inches,
       Octagonal Scales, ... $1\frac{1}{2}$ "   "   11.00

BOXWOOD RULES.

Two Feet.   Two Fold.

No. 29.   One-Third Size.

No.		Per Dozen
29.	Round Joint, 8ths and 16ths of inches, ... 1½ in. wide	$2.75

No. 18.   One-Third Size.

18.   Square Joint, 8ths and 16ths of inches, ... 1½ in. wide   4.00
1.   Arch Joint, 8ths and 16ths of inches and Octagonal
         Scales, ... 1½ "   "   4.50

No. 2.   One-Third Size.

2.   Arch Joint, 8ths, 10ths, and 16ths of inches, Octag-
         onal Scales, ... 1½ in. wide   5.00
3½.   Arch Joint, *Extra Thin,* 8ths and 16ths of inches,
         Drafting and Octagonal Scales, ... 1½ "   "   6.50

No. 4.   One-Third Size.

4.   Arch Joint, *Extra Thin,* Plates on outside of wood,
         8ths and 16ths of inches, Drafting and Octag-
         onal Scales, ... 1½ in. wide   6.50
5.   Arch Joint, Bound, 8ths, 10ths and 16ths of inches,
         Drafting and Octagonal Scales, ... 1½ "   "   9.50

BOXWOOD RULES.

SLIDE.

Two Feet.   **Two Fold.**

No.		Per Dozen
26.	Square Joint, Plain Slide, 8ths, 10ths, and 16ths of inches, ... 1½ in. wide	$7.00
27.	Square Joint, Gunter Slide, 8ths, 10ths, and 16ths of inches, 100ths of feet, ... 1½ " "	10.00
11.	Arch Joint, Plain Slide, 8ths, 10ths, and 16ths of inches, ... 1½ " "	8.00

No. 12. One-Third Size.

12.	Arch Joint, Gunter Slide, 8ths, 10ths, and 16ths of inches, 100ths of feet, ... 1½ in. wide	11.00
15.	Arch Joint, Bound, Gunter Slide, 8ths, 10ths and 16ths of inches, ... 1½ " "	15.00
6.	Arch Joint, Engineers' Scales, Gunter Slide, 8ths, 10ths, and 16ths of inches, 100ths of feet, . 1½ " "	20.00
16.	Arch Joint, Bound, Engineers' Scales, Gunter Slide, 8ths, 10ths, and 16ths of inches, . . 1½ " "	24.00

Books of Instruction for Engineers, 25 cents each, Net.

EXTRA NARROW.

Two Feet.   **Four Fold.**

61½.   Square Joint, Middle Plates, 8ths and 16ths of in.,   ¾ in. wide   3.25

No. 63½. Two-Thirds Size.

63½.	Square Joint, Edge Plates, 8ths, 10ths, and 16ths of inches, ... ¾ in. wide	4.25
53¼.	Arch Joint, Edge Plates, 8ths, 10ths, and 16ths of inches, ... ¾ " "	5.00
62½.	Square Joint, Bound, 8ths, 10ths, 12ths, and 16ths of inches, ... ¾ " "	8.00

BOXWOOD RULES

Architects.   Inner Edges Beveled.

Two Feet.   Four Fold.

EXTRA NARROW.

No.   Per Dozen

53¾.   Arch Joint, Edge Plates, 8ths, 10ths, 12ths, and
16ths of inches, with Drafting Scales,   .   .   ¾ in. wide $9.00

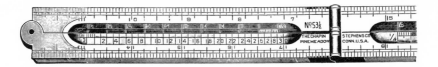

No. 53½.   One-Third Size.

NARROW.

53½.   Arch Joint, Edge Plates, 8ths, 10ths, 12ths, and
16ths of inches, with Drafting Scales,   .   .   1 in. wide   8.00

No. 53½ Graduated on outside edges in 100ths of a foot, 50c. per dozen, net extra.

BROAD.

75½.   Arch Joint, Edge Plates, 8ths, 10ths, 12ths and 16ths
of inches, with Drafting and Octagonal Scales, 1⅜ in. wide   10.00

SHIP CARPENTERS' BEVELS.

No. 42.   Two-Thirds Size.

42.   Boxwood, Double Tongue, 8ths and 16ths, ...   4.00
43.   Boxwood, Single Tongue, 8ths and 16ths, ...   4.00

BOXWOOD RULES.

CALIPER.   **TWO FOLD.**

Slides in 32nds of Inches.

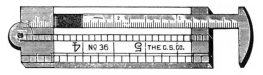

No. 36. Two-Thirds Size.

No.   Per Dozen

36.	Square Joint, 6-inch, 8ths, 10ths, 12ths, and 16ths of inches, ...	$\frac{7}{8}$ in. wide	$4.50
13.	Square Joint, 6-inch, 8ths and 16ths of inches, . .	$1\frac{1}{8}$ " "	5.50
13½.	Square Joint, 6-inch, 8ths and 16ths of inches, . .	$1\frac{1}{2}$ " "	6.50

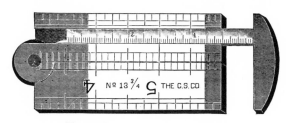

No. 13¾. Two-Thirds Size.

13¾.	Square Joint, 6-inch, 8ths, 10ths, 12ths, and 16ths of inches, 3-inch Slide running through Head,	$1\frac{1}{2}$ in. wide	8.00
14.	Square Joint, 6-inch, Brass Case, 16ths of inches, .	$\frac{7}{8}$ " "	8.00
14½.	Square Joint, Bound, 6-inch, 8ths, 10ths, 12ths, and 16ths of inches, ...	$\frac{7}{8}$ " "	7.00

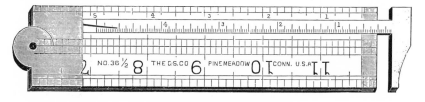

No. 36½. Two-Thirds Size.

36½.	Square Joint, 12-inch, 8ths, 10ths, 12ths, and 16ths of inches, ...	$1\frac{3}{8}$ in. wide	6.50

Caliper Rules made with slides right hand, 25c per dozen net extra.

BOXWOOD RULES.

CALIPER.   FOUR FOLD.

Slides in 32nds of Inches.

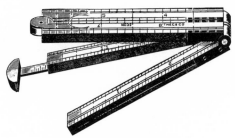

No. 32.   Two-Thirds Size.

No.   Per Dozen

32.   Arch Joint, Edge Plates, 12-inch, 8ths, 10ths, 12ths,
                and 16ths of inches, ... $\frac{7}{8}$ in. wide $7.00

32½.   Arch Joint, Bound, 12-inch, 8ths, 10ths, 12ths, and
                16ths of inches, ... $\frac{7}{8}$ "   "   10.00

No. 3.   Two-Thirds Size.

3.   Square Joint, Bound, 12-inch, 8ths and 16ths of
                inches, ... $\frac{5}{8}$ in. wide   12.00

62C.   Square Joint, Two Feet, 8ths, 10ths, 12ths, and
                16ths of inches, ... 1 "   "   14.00

83½.   Arch Joint, Edge Plates, Two Feet, 8ths, 10ths, and
                16ths of inches and Scales, ... 1$\frac{3}{8}$ "   "   12.00

No. 76C.   One-Third Size.

76C.   Arch Joint, Full Bound, Two Feet, 8ths and 16ths
                of inches, Drafting and Octagonal Scales,   1$\frac{3}{8}$ in. wide   16.00

Caliper Rules made with slides right hand, 25c per dozen net extra.

BOXWOOD RULES.

NARROW.

Three Feet.   Four Fold.

No.   Per Dozen

68¼. Round Joint, Middle Plates, 8ths and 16ths of inches, 1 in. wide $4.50
61¼. Square Joint, Middle Plates, 8ths and 16ths of inches, 1 "   "   5.25

No. 66.   One-Half Size.

66.   Arch Joint, Middle Plates, Yard Division outside and
16ths of inches, inside, ... 1 in. wide   6.00

No. 66½.   One-Half Size.

66½. Arch Joint, Middle Plates, 8ths of inches outside
and 16ths of inches, inside, ... 1 in. wide   6.00
66¼. Arch Joint, Bound, Yard Divisions outside and
16ths of inches inside, ... 1 "   "   15.00
66¾. Arch Joint, Bound, 8ths of inches outside and 16ths
of inches, inside, ... 1 "   "   15.00

BROAD.

Three Feet.   Four Fold.

67¼. Round Joint, Middle Plates, 8ths and 16ths of inches, 1⅜ in. wide   5.50
70¼. Square Joint, Middle Plates, 8ths and 16ths of inches, 1⅜ "   "   6.00
73¼. Arch Joint, Middle Plates, 8ths and 16ths of inches, 1⅜ "   "   8.00
77¼. Double Arch Joint, 8ths and 16ths of inches, . . 1⅜ "   "   10.00

BOXWOOD RULES.
Two Feet.   Six Fold.

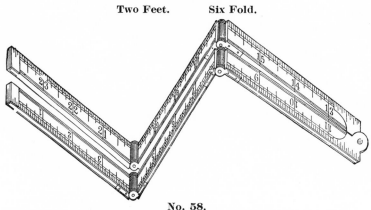

No. 58.

<table>
<tr><td>No.</td><td></td><td></td><td>Per Dozen</td></tr>
<tr><td>58.</td><td>Arch Joint, Edge Plates, 8ths, 10ths, 12ths, and 16ths of inches, ...</td><td>¾ in. wide</td><td>$6.50</td></tr>
<tr><td>58½.</td><td>Arch Joint, Bound, 8ths, 10ths, 12ths, and 16ths of inches, ...</td><td>¾ " "</td><td>18.00</td></tr>
</table>

One Foot.   Three Fold.

58¾.   Edge Plates, 8ths and 16ths of inches, ...   ½ " "   3.00

CARRIAGE MAKERS' RULES.
Four Feet.   Four Fold.

No. 94.   One-Third Size.

94.   Arch Joint, Bound, 8ths and 16ths of inches, . .   1½ in. wide   26.00

WINDOW RULE.
Four Feet.   Three Fold.

No. 98.   One-Third Size.

98.   Eights of inches, sliding from two to four feet, . .   ⅞ in. wide   36.00

Boxwood Rules marked left to right (English style) and Boxwood and Ivory Rules graduated in Metric or Spanish division to order.

IVORY RULES.

POCKET.

One Foot.   Four Fold.

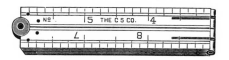

No. 90.   Two-Thirds Size.

No.   Per Dozen

90.   Round Joint, Middle Plates, Brass, 8ths and 16ths
 of inches, ...   ½ in. wide $10.00

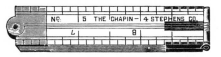

No. 90½.   Two-Thirds Size.

90½.   Square Joint, Middle Plates, Brass, 8ths and 16ths
 of inches, ...   ½ in. wide   12.00

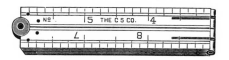

No. 91.   Two-Thirds Size.

91.   Square Joint, Edge Plates, German Silver, 8ths,
 10ths, 12ths, and 16ths of inches, ...   ⅞ in. wide   23.00
91½.   Square Joint, Bound, German Silver, 8ths and
 16ths of inches, ...   ⅝ "   "   28.00
92.   Square Joint, Edge Plates, German Silver, 8ths and
 16ths of inches, ...   ⅝ "   "   17.00
92½.   Square Joint, Middle Plates, German Silver, 8ths
 and 16ths of inches, ...   ½ "   "   14.00

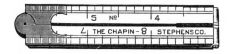

No. 88.   Two-Thirds Size.

88.   Arch Joint, Bound, German Silver, 8ths and 16ths
 of inches, ...   ⅝ in. wide   32.00
88½.   Arch Joint, Edge Plates, German Silver, 8ths and
 16ths of inches, ...   ⅝ "   "   21.00

IVORY RULES.
TWO FOLD.

98½. Round Joint, 6-inch, Brass, 8ths and 16ths of inches, ½ in. wide   4.50
99½. Arch Joint, 12-inch, German Silver, 8ths, 10ths,
12ths, and 16ths of inches, ...   ¾ "   "   24.00

NARROW.
Two Feet.   Four Fold.

No.   Per Dozen
85. Square Joint, Edge Plates, German Silver, 8ths,
10ths, 12ths, and 16ths of inches,   . . .   ⅞ in. wide $54.00
86. Arch Joint, Edge Plates, German Silver, 8ths,
10ths, 12ths, and 16ths of inches, 100ths of
feet, Drafting Scales, ...   1 "   "   64.00
87. Arch Joint, Bound, German Silver, 8ths, 10ths,
12ths, and 16ths of inches, Drafting Scales,   1 "   "   80.00
89. Double Arch Joint, Bound, German Silver, 8ths,
10ths, 12ths, and 16ths of inches, Drafting
Scales, ...   1 "   "   92.00

BROAD.
Two Feet.   Four Fold.

94½. Arch Joint, Edge Plates, German Silver, 8ths,
10ths, 12ths, and 16ths of inches, Drafting
Scales, ...   1⅜ in wide   84.00
95. Arch Joint, Bound, German Silver, 8ths, 10ths,
12ths, and 16ths of inches, Drafting Scales,   1⅜ "   "   102.00
95½. Arch Joint, Bound, with *Slide,* German Silver,
8ths, 10ths, 12ths, and 16ths of inches, Drafting Scales, ...   1⅜ "   "   112.00
95¾. Arch Joint, Bound, with *Caliper,* German Silver,
8ths, 10ths, 12ths, and 16ths of inches, Drafting Scales, ...   1⅜ "   "   120.00
97. Double Arch Joint, Bound, German Silver, 8ths,
10ths, 12ths, and 16ths of inches, Drafting
Scales, ...   1⅜ "   "   116.00

ARCHITECTS'.
With Inside Edges Beveled.
EXTRA NARROW.
Two Feet.   Four Fold.

85½. Arch Joint, German Silver Edge Plates, 8ths, 10ths,
12ths, and 16ths, with Drafting Scales,   . .   ¾ in. wide   96.00

NARROW.
Two Feet.   Four Fold.

86½. Arch Joint, German Silver Edge Plates, 8ths, 10ths,
12ths, and 16ths, with Drafting Scales,   . .   1 in. wide   96.00

IVORY RULES.

CALIPER.

Six Inch.   Two Fold.

Slides in 32nds of Inches.

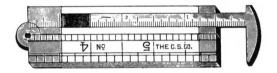

No. 38.   Two-Thirds Size.

No.   Per Dozen

38.   Square Joint, German Silver, 8ths, 10ths, 12ths, and
         16ths of inches, ... $\frac{7}{8}$ in. wide $15.00
40¼. Square Joint, German Silver, Bound, 8ths, 10ths,
         12ths, and 16ths of inches, ... $\frac{7}{8}$ "   "   30.00

No. 40½.   Two-Thirds Size.

40½. Square Joint, German Silver, Bound, 8ths and 16ths
         of inches, ... $\frac{5}{8}$ in. wide   24.00
40¾. Square Joint, German Silver Case, 16ths of inches, $\frac{7}{8}$ "   "   22.00

One Foot.   Four Fold.

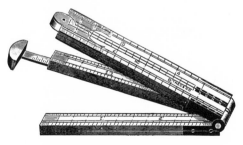

No. 39.   Two-Thirds Size.

39.   Square Joint, Edge Plates, German Silver, 8ths,
         10ths, 12ths, and 16ths of inches, ... $\frac{7}{8}$ in. wide   38.00

IVORY RULES.

CALIPER.

One Foot.   Four Fold.

Slides in 32nds of Inches.

No. 39½.   Two-Thirds Size.

No.
                                             Per Dozen
39½. Square Joint, Bound, German Silver, 8ths, 10ths,
       12ths, and 16ths of inches, ...   ⅞ in. wide $45.00

No. 40.   Two-Thirds Size.

40.   Square Joint, Bound, German Silver, 8ths and 16ths
       of inches, ...   ⅝ in. wide 44.00

No. 99¾.   Two-Thirds Size.

99¾. Arch Joint, Bound, German Silver, 8ths and 16ths
       of inches, ...   ⅝ in. wide 46.00
99¼. Arch Joint, Bound, German Silver, 8ths, 10ths,
       12ths, and 16ths of inches, ...   ⅞ "   "   48.00

STEPHENS' PATENT COMBINATION RULE.

12 Inch.   Two Fold.

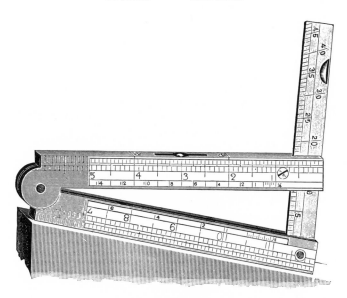

No. 036.   Cut One-Half Size.

Price, ...   per dozen, $40.00

The engraving illustrates an instrument invented by L. C. Stephens, and patented by him, which combines in itself a Carpenter's Rule, Spirit Level, Square, Plumb, Bevel, Inclinometer or Slope Level, Brace Scale, Draughting Scale, T Square, Protractor, Right-Angled Triangle, and with a straight edge can be used as a parallel Ruler.

This Combination Rule is made of boxwood, with one joint, and is well protected with heavy brass binding. The plate which protects the glass being put on with screws, can be removed should it by accident become necessary to insert a new glass.

When folded it is six inches long, one and three-eighths inches wide, and three-eighths of an inch thick, and weighs the same as an ordinary broad-bound rule. The cut represents the rule in use as a clinometer, or *Slope-Level;* in which it is represented in taking the angle or inclination of an inclined plane— the top of a desk for instance. The steel blade folds like a knife blade into the part which holds it.

A favorite instrument with miners as a Slope Level.

Carpenters, joiners, ship builders, draughtsmen, engineers, roofers and all classes of mechanics are unanimous in the approval of this device and the symmetrical arrangements of its parts.

FLEXIFOLD RULES.

WITH RIVET SPRING JOINTS.

Strongest and best Flexible Folding Rules made.   In Yellow and White Enamel.   Packed ½ Dozen in a Red Box.   Look for trade mark "Flexifold" on Rules and Boxes.

Graduated 16th inches both sides.

	YELLOW.			WHITE.	
Length	No.	Per Doz.	Length	No.	Per Doz.
2 feet	220	$2.00	2 feet	222	$2.28
3 "	330	3.00	3 "	333	3.36
4 "	440	4.00	4 "	444	4.44
5 "	550	5.00	5 "	555	5.52
6 "	660	6.00	6 "	666	6.60
8 "	880	8.00	8 "	888	8.88

Graduated Metric on one side and regular 16th inches on the other.

	YELLOW.			WHITE.	
Length	No.	Per Doz.	Length	No.	Per Doz.
3 feet	330M	$3.00	3 feet	333M	$3.36
4 "	440M	4.00	4 "	444M	4.44
5 "	550M	5.00	5 "	555M	5.52
6 "	660M	6.00	6 "	666M	6.60

"PEARCE" RULES.

WITH RIVET SPRING JOINTS.

All "Pearce" Rules packed One Dozen in a Red Box.   Look for trade mark "Pearce' on Rules and Boxes.

Graduated 16th inches both sides.

YELLOW.			WHITE.		
Length	No.	Per Doz.	Length	No.	Per Doz.
2 feet	212	$1.80	2 feet	232	$2.04
3 "	213	2.70	3 "	233	3.00
4 "	214	3.60	4 "	234	3.96
5 "	215	4.50	5 "	235	4.92
6 "	216	5.40	6 "	236	6.00
8 "	218	7.20	8 "	238	7.92

Graduated Metric on one side and regular 16th inches on the other.

YELLOW.			WHITE.		
Length	No.	Per Doz.	Length	No.	Per Doz.
3 feet	213M	$2.70	3 feet	233M	$3.00
4 "	214M	3.60	4 "	234M	3.96
5 "	215M	4.50	5 "	235M	4.92
6 "	216M	5.40	6 "	236M	6.00

STATIONERS' GOODS.

DESK RULERS.

Paper Cutting Edge.   Beech and Maple Polished.

ONE DOZEN IN A BOX.

Graduated 8ths.

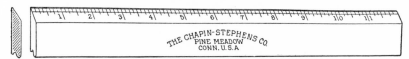

Plain Edge.

Length		Width	Per Gross
12 inch	. . . .	1⅜ inch	$9.00
15 inch	. . . .	1⅜ inch	11.00
18 inch	. . . .	1⅜ inch	13.00
21 inch	. . . .	1⅜ inch	15.00
24 inch	. . . .	1⅜ inch	18.00

Brass Edge.

Length		Width	Per Gross
12 inch	. . . .	1⅜ inch	$18.00
15 inch	. . . .	1⅜ inch	22.00
18 inch	. . . .	1⅜ inch	27.00
21 inch	. . . .	1⅜ inch	31.00
24 inch	. . . .	1⅜ inch	36.00

BOXWOOD RULERS.

Graduated 16ths.

Brass Edge.

Length		Width	Per Gross
12 inch	. . . .	1⅜ inch	$41.00
15 inch	. . . .	1⅜ inch	46.00
18 inch	. . . .	1⅜ inch	54.00
21 inch	. . . .	1⅜ inch	64.00
24 inch	. . . .	1⅜ inch	77.00

STEPHENS' PATENT DESK RULERS.

Maple, Brass Edge.   Colored Grooves.

Graduated 8ths.

Length		Width	Per Gross
12 inch	. . . .	1⅜ inch	$25.50
15 inch	. . . .	1⅜ inch	32.00
18 inch	. . . .	1⅜ inch	38.00
21 inch	. . . .	1⅜ inch	46.00
24 inch	. . . .	1⅜ inch	56.00

Boxwood, Brass Edge.   Grooves Not Colored.

Graduated 16ths.

Length		Width	Per Gross
12 inch	. . . .	1⅜ inch	$59.00
15 inch	. . . .	1⅜ inch	74.00
18 inch	. . . .	1⅜ inch	89.00
21 inch	. . . .	1⅜ inch	106.00
24 inch	. . . .	1⅜ inch	118.00

SCHOOL RULERS.

Graduated 8ths.

Length		Width	Per Gross
Beech, 12 inch,	. . . .	¾ inch	$4.00
Boxwood, 12 inch,	. . . .	¾ inch	16.00
Beech, Chicago Pattern, 12-inch,	. . . . . .	1⅛ inch	5.00

MISCELLANEOUS RULES.

PATTERN MAKERS' SHRINKAGE RULES.

No. 30.

No.			Per Dozen
30.	Boxwood, 24¼ inches long, ...	1½ in. wide	$15.00

BENCH RULES.

34.	Maple, Brass Tips, 24-inch, ...	1¼ "	"	4.00
35.	Maple, with Board Measure Tables, 24-inch, . .	1½ "	"	6.50
35¼.	Boxwood, 16ths inches, 24-inch, ...	1½ "	"	13.00

YARD STICKS.

33½.	Plain, ...	1 "	"	1.50
33.	Polished, ...	1 "	"	2.00

No. 41.

41.	Brass Tipped, Polished, ...	1 in. wide	3.50

No. 41½.

41½.	Brass Tipped, Polished, ½ inch square, . . .		4.00
50.	Hickory, Brass Capped, Polished, Flat, . . .	¾ in. wide	4.50

SADDLER'S RULE.

80.	Maple, 36-inch, Brass Tips, ...	1½ in. wide	9.00

MISCELLANEOUS RULES.

WANTAGE AND GAUGING RODS.

No.			Per Dozen
44.	Wantage Rod, 8 Lines, ...		$7.00
37.	Wantage Rod, 12 Lines, ...		10.00
45.	Gauging Rod, ...	36 inch,	7.00
45¼.	Gauging Rod, ...	48 inch,	8.00
45½.	Gauging Rod with Wantage Table, ...	48 inch,	18.00

BOARD MEASURES.

EXPLANATION OF BOARD STICKS.—Know the length of boards you wish to measure. The figures on the end, eight and upwards, is the length in feet; place the Stick on the flat surface to the outer edge of the board, follow the length column to the opposite edge, and the figure on the edge will be the contents in feet of 1 inch boards.

No.			Per Dozen
45¾.	Board Stick, Octagon, 8 Lines, 9 to 16 feet, . . .	24 inch,	$15.00
46.	Board Stick, Octagon, 16 Lines, 8 to 23 feet, . . .	24 inch,	16.00
46½.	Board Stick, Square, 16 Lines, 8 to 23 feet, . . .	24 inch,	16.00
47.	Board Stick, Octagon, 16 Lines, 8 to 23 feet, . . .	36 inch,	26.00
47½.	Board Stick, Square, 16 Lines, 8 to 23 feet, . . .	36 inch,	26.00
48.	Board Stick, Walking Cane, Brass Head and Tip, 8 Lines, 9 to 16 feet, Octagon, ...	36 inch,	21.00
43¼.	Board Stick, Flat, with T Head, 10 Lines, 9 to 19 feet,	36 inch,	12.00
43½.	Board Stick, Hickory, Flat, with T Head, 8 Lines, 9 to 16 feet, ...	36 inch,	15.00
49.	Board Stick, Flat Hickory, Extra Thin, Steel Head, Extra Strong, 6 Lines, 12 to 22 feet, . . .	36 inch,	22.00

LOG MEASURES.

EXPLANATION OF LOG STICKS.—These Sticks give the number of feet of 1 inch square edge boards sawed from a log from 12 to 36 inches in diameter. The figures 12 to 20, near the head, are for the lengths of logs in feet; follow the column under the length of the log to the diameter of the log, which will give the number of feet the log will make. Logs not over 15 feet long, the diameter should be taken at the small end; over 15 feet in length, at the middle.

No.			Per Dozen
48¾.	Log Stick, Flat, with T Head, ...	36 inch,	$16.00
48¼.	Log Stick, Hickory, Flat, with T Head, ...	36 inch,	18.50
48½.	Log Stick, Walking Cane, Brass Head and Tip, . .	36 inch,	26.00

COMMON BENCH PLANES.

Stamped J. Pearce.

COMMON BENCH PLANES, SINGLE IRONS.

WITH IRON STARTS.

No.			Price, Each
100.	Smooth, Best Cast Steel Irons, ...	1¾ to 2¼ inch,	$0.70
101.	Jack, 16-inch, Best Cast Steel Irons, . . .	2 to 2¼ inch,	.85
102.	Fore, 18 to 22-inch, Best Cast Steel Irons, .	2⅜ to 2½ inch,	1.40
103.	Jointer, 26-inch, Best Cast Steel Irons, . .	2½ inch,	1.50
	Jointer, 28-inch, Best Cast Steel Irons, . .	2½ inch,	1.60
	Jointer, 30-inch, Best Cast Steel Irons, . .	2½ inch,	1.80

COMMON BENCH PLANES, DOUBLE IRONS.

WITH IRON STARTS.

No.			Price, Each
108.	Smooth, Best Cast Steel Irons, ...	1¾ to 2¼ inch,	$0.90
108½.	Smooth, Jack Handle, Best Cast Steel Irons,	2 to 2¼ inch,	1.50
109.	Jack, 16-inch, Best Cast Steel Irons, . . .	2 to 2¼ inch,	1.00
109½.	Jack, Razee Handle, Best Cast Steel Irons,	2 to 2¼ inch,	1.20
110.	Fore, 18 to 22-inch, Best Cast Steel Irons, .	2⅜ to 2⅝ inch,	1.70
111.	Jointer, 26-inch, Best Cast Steel Irons, . .	2½ to 2¾ inch,	1.80
	Jointer, 28-inch, Best Cast Steel Irons, . .	2½ to 2¾ inch,	1.90
	Jointer, 30-inch, Best Cast Steel Irons, . .	2½ to 2¾ inch,	2.15

BENCH PLANES, WITHOUT IRONS.

No.		Price, Each
100Z.	Smooth, Single, ...	$0.60
101Z.	Jack, Single, ...	.70
102Z.	Fore, Single, ...	1.30
103Z.	Jointer, 26-inch, Single, ...	1.40
	Jointer, 28-inch, Single, ...	1.50
	Jointer, 30-inch, Single, ...	1.70
108Z.	Smooth, Double, ...	.60
109Z.	Jack, Double, ...	.70
110Z.	Fore, Double, ...	1.30
111Z.	Jointer, 26-inch, Double, ...	1.40
	Jointer, 28-inch, Double, ...	1.50
	Jointer, 30-inch, Double, ...	1.70

EXTRA BENCH PLANES.

No. 104.

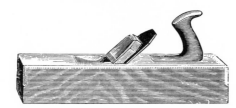

No. 105.

No. 106.   22 Inch.

EXTRA BENCH PLANES, SINGLE IRONS.

WITH IRON OR WOOD STARTS, IF REQUESTED.

No.			Price, Each
104.	Smooth, Best Cast Steel Irons, ...	1¾ to 2¼ inch,	$0.70
105.	Jack, 16-inch, Best Cast Steel Irons, . . .	2 to 2¼ inch,	.85
105½.	Jack, Razee Handle, Best Cast Steel Irons, .	2 to 2½ inch,	1.10
106.	Fore, 18 to 22-inch, Best Cast Steel Irons, .	2⅜ to 2½ inch,	1.40
106½.	Fore Razee Handle, 18 to 22-inch, Best Cast Steel Irons, ...	2⅜ to 2½ inch,	1.70
107.	Jointer, 24 to 26-inch, Best Cast Steel Irons, . .	2½ inch,	1.50
	Jointer, 28-inch, Best Cast Steel Irons, ...	2½ inch,	1.60
	Jointer, 30-inch, Best Cast Steel Irons, ...	2½ inch,	1.80
107½.	Jointer, Razee Handle, 24 to 26-inch, Best Cast Steel Irons, ...	2½ inch,	1.80
	Jointer, Razee Handle, 28-inch, Best Cast Steel Irons,	2½ inch,	1.90
	Jointer, Razee Handle, 30-inch, Best Cast Steel Irons,	2½ inch,	2.10
	Planes with extra sized Single Irons, per ⅛ inch, add to List Price,		.10
	Planes with English Irons, Single, add to List Price, ...		.15
	Planes with Handle Bolts, add to List Price, ...		.25

EXTRA BENCH PLANES.

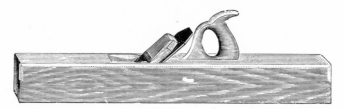

No. 115.   26 Inch.

No. 112¾.   **No. 112½.**

EXTRA BENCH PLANES, DOUBLE IRONS.

WITH IRON OR WOOD STARTS, IF REQUESTED.

No.			Price, Each
112.	Smooth, Best Cast Steel Irons, ...	1¾ to 2¼ inch,	$0.90
112½.	Smooth, Solid Handle, Best Cast Steel Irons, .	2 to 2¼ inch,	1.75
112¾.	Smooth, Solid Handle, Improved, Best Cast Steel Irons, ...	2 to 2¼ inch,	1.75
113.	Jack, 16-inch, Best Cast Steel Irons, . . .	2 to 2¼ inch,	1.00
113½.	Jack, Razee Handle, Best Cast Steel Irons, .	2 to 2¼ inch,	1.20
114.	Fore, 18 to 22-inch, Best Cast Steel Irons, .	2⅜ to 2⅝ inch,	1.70
114½.	Fore, Razee Handle, 18 to 22-inch, Best Cast Steel Irons, ...	2⅜ to 2⅝ inch,	1.90
115.	Jointer, 24 to 26-inch, Best Cast Steel Irons, .	2½ to 2¾ inch,	1.80
	Jointer, 28-inch, Best Cast Steel Irons, . .	2½ to 2¾ inch,	1.90
	Jointer, 30-inch, Best Cast Steel Irons, . .	2½ to 2¾ inch,	2.15
115½.	Jointer, Razee Handle, 24 to 26-inch, Best Cast Steel Irons, ...	2½ to 2¾ inch,	2.05
	Jointer, Razee Handle, 28-inch, Best Cast Steel Irons, ...	2½ to 2¾ inch,	2.15
	Jointer, Razee Handle, 30-inch, Best Cast Steel Irons, ...	2½ to 2¾ inch,	2.40

Planes with extra sized Double Irons, per ⅛ inch, add to List Price, ...15

Planes with English Irons, double, add to List Price, ...30

Planes with Handle Bolt, add to List Price, ...25

PREMIUM BENCH PLANES.

EXTRA CAST STEEL DOUBLE IRONS.

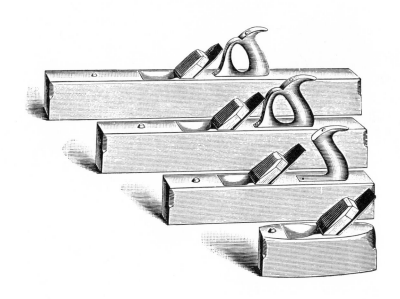

With Bolted Handle and Start.

No.		Price, Each
400.	Smooth, ...	$1.00
401.	Jack, 16-inch, ...	1.25
401½.	Jack, 16-inch, Razee, ...	1.45
402.	Fore, 22-inch, ...	2.30
402½.	Fore, 22-inch, Razee, ...	2.50
403.	Jointer, 26-inch, ...	2.60
	Jointer, 28-inch, ...	2.75
	Jointer, 30-inch, ...	2.95
403½.	Jointer, 26-inch, Razee, ...	2.80
	Jointer, 28-inch, Razee, ...	2.95
	Jointer, 30-inch, Razee, ...	3.15

Ebony or Boxwood Starts in place of Iron if requested.

PREMIUM BENCH PLANES.

Polished.

EXTRA CAST STEEL DOUBLE IRONS.

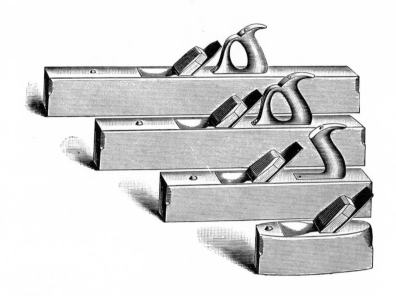

With Bolted Handle and Start.

No.		Price, Each
404.	Smooth, ...	$1.50
405.	Jack, 16-inch, ...	1.85
405½.	Jack, 16-inch, Razee, ...	2.05
406.	Fore, 22-inch, ...	2.55
406½.	Fore, 22-inch, Razee, ...	2.75
407.	Jointer, 26-inch, ...	2.90
407½.	Jointer, 26-inch, Razee, ...	3.10
408.	Smooth, Solid Handle, ...	2.25
409.	Mitre, Smooth Shape, Single Iron, ...	1.00
410.	Mitre, Smooth Shape, Double Iron, ...	1.25

Ebony or Boxwood Starts in place of iron, if requested.

SHIP PLANES.

No. 424.

No. 425.

No. 427.

No.		Price, Each
423.	Smooth, 9-inch, Best Cast Steel Double Irons, . to 1¾ inch,	$1.00
424.	Jack, 16-inch, Razee, Best Cast Steel Double Irons, to 1⅞ inch,	1.20
425.	Fore, 22-inch, Razee, Best Cast Steel Double Irons, to 2⅛ inch,	1.90
426.	Jointer, 26-in., Razee, Best Cast Steel Double Irons, to 2¼ inch,	2.10
427.	Spar Plane, Best Cast Steel Double Irons, ...	1.50
428.	Spar Plane, Best Cast Steel Single Irons, ...	1.00

When ordering Numbers 427 or 428 give size of spar to be worked.

MISCELLANEOUS PLANES.

No. 431.

No. 432.

No. 430.

No.			Price, Each
430.	Tooth Plane, Single Iron,	1⅝ to 2¼ inch,	$1.00
431.	Mitre Plane, Square, Single Iron,	1½ to 1¾ inch,	.75
432.	Mitre Plane, Smooth Shape, Single Iron,	1½ to 1¾ inch,	.75
433.	Mitre Plane, Square, Double Iron,	1¾ inch,	1.00
434.	Mitre Plane, Smooth Shape, Double Iron,	1¾ inch,	1.00
436.	Compass or Circular Smooth, Single Iron,	2 to 2⅛ inch,	1.00
437.	Compass or Circular Smooth, Double Iron	2 to 2⅛ inch,	1.25

No. 438.

No. 440.

438.	Toy, Smooth, Single Iron,	1½ inch,	$0.65
439.	Toy, Smooth, Double Iron,	1½ inch,	.75
440.	Toy, Jack, Single Iron,	1½ inch,	.75
441.	Toy, Jack, Double Iron,	1½ inch,	.85

No. 442.

No. 448.

442.	Gutter Plane (give size of circle for face),	1½ to 2 inch,	$1.25
447.	German Smooth or Bull Plane, Single Iron,		1.20
448.	German Smooth or Bull Plane, Double Iron,		1.25
449.	Box Maker's Jack, Razee, Single Iron,	2 to 2¼ inch,	1.75

APPLEWOOD, BOXWOOD, AND ROSEWOOD PLANES.

POLISHED.

With Iron or Wood Starts.   Extra C. S. Double Irons.

No.			Price, Each
411.	Smooth, Applewood, ...	2 to 2¼ inch,	$1.50
411½.	Smooth, Applewood, Solid Handle, ...	2 to 2¼ inch,	2.25
412.	Jack, Applewood, Bolted Handle, ...	2 to 2¼ inch,	1.75
412½.	Jack, Applewood, Razee, Bolted Handle, . .	2 to 2¼ inch,	2.25
415.	Smooth, Boxwood, ...	2 to 2¼ inch,	2.50
415½.	Smooth, Boxwood, Small Extra, ...	1¾ inch,	1.75
417.	Smooth, Rosewood, ...	2 to 2¼ inch,	2.50
417½.	Smooth, Rosewood, Small Extra, ...	1¾ inch,	2.00
419.	Mitre, Boxwood, Single Iron, ...	1½ to 1¾ inch,	2.00
420.	Mitre, Boxwood, Double Iron, ...	1½ to 1¾ inch,	2.25
421.	Mitre, Rosewood, Single Iron, ...	1½ to 1¾ inch,	1.75
422.	Mitre, Rosewood, Double Iron, ...	1½ to 1¾ inch,	2.00

CARRIAGE MAKERS' TOOLS.

No. 1.   **No. 2.**

No.   Price, Each
1. Carriage Makers' Smooth, Double Iron, ... to 1⅝ inch, $0.90
2. Carriage Makers' Smooth, Circle Face, Double Iron, to 1⅝ inch, 1.05

No. 3.   **No. 5.**

3. Carriage Makers' Rabbet Plane, ...   1   inch, $1.00
4. Carriage Makers' Rabbet Plane, Circle Face, .   .   1   inch, 1.10
5. Carriage Makers' T Rabbet Plane, ... to 1½ inch, 1.20
6. Carriage Makers' T Rabbet Plane, Circle Face,   . to 1½ inch, 1.35

No. 7.

7. Carriage Makers' Beading Tool, ... 1.75

No. 9.

8. Carriage Makers' Router, Single Cutter, ... 1.00
9. Carriage Makers' Router, Double Cutter, ... 1.20

No. 10.

10. Carriage Makers' Router, Double Cutter, with Guard, ... 1.40
11. Carriage Makers' Panel Router, ... 5.00
12. Carriage Makers' Boxing Tool, ... 1.60

BEAD PLANES—BEADS.

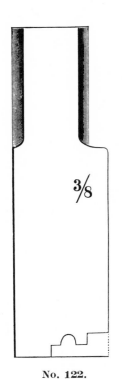

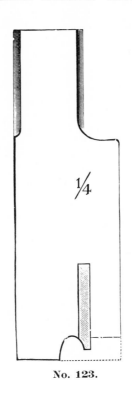

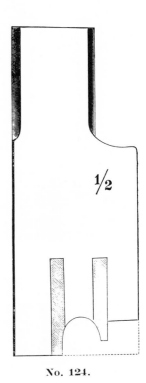

No. 122.   No. 123.   No. 124.

No.		Price, Each
122.	Astragals, ... ¼, ⅜, ½, ⅝, ¾, ⅞, 1 inch,	$0.55
	Astragals, ... 1⅛, 1¼ inch,	.65
	Astragals, ... 1⅜, 1½ inch,	.75
123.	Beads, Single Boxed, . . . ⅛, $\frac{3}{16}$, ¼, $\frac{5}{16}$, ⅜, $\frac{7}{16}$, ½ inch,	.50
	Beads, Single Boxed, ... ⅝, ¾ inch,	.55
	Beads, Single Boxed, ... ⅞, 1 inch,	.70
	Beads, Single Boxed, ... 1⅛, 1¼ inch,	.90
	Beads, Single Boxed, ... 1⅜, 1½ inch,	1.00
	Beads, Single Boxed, with Handle, ... 1½ inch,	1.90
124.	Beads, Double Boxed, . . . ⅛, $\frac{3}{16}$, ¼, $\frac{5}{16}$, ⅜, $\frac{7}{16}$, ½ inch,	.60
	Beads, Double Boxed, ... ⅝, ¾ inch,	.65
	Beads, Double Boxed, ... ⅞, 1 inch,	.80
	Beads, Double Boxed, ... 1⅛, 1¼ inch,	1.00
	Beads, Double Boxed, ... 1½ inch,	1.10

Beads, Double, Right and Left Hand, in one Block, Double Price.

BEAD PLANES—BEADS.

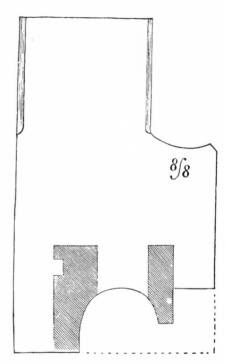

No. 125.   No. 125.

125. Beads, Solid Boxed, Dovetailed, . . .   ⅛, ³⁄₁₆, ¼, ⁵⁄₁₆ inch, $0.70

Beads, Solid Boxed, Dovetailed,. . . . . ⅜, ⁷⁄₁₆, ½ inch,   .80

Beads, Solid Boxed, Dovetailed, ... ⅝, ¾ inch,   .90

Beads, Solid Boxed, Dovetailed, ... ⅞, 1 inch, 1.00

Beads, Solid Boxed, Dovetailed, ... 1⅛, 1¼ inch, 1.20

Beads, Double, Right and Left Hand, in one Block, Double Price.

Beads with Solid Handle, add to List Price, ... 1.00

BEADS AND REEDING PLANES.

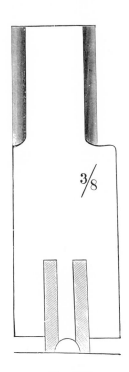

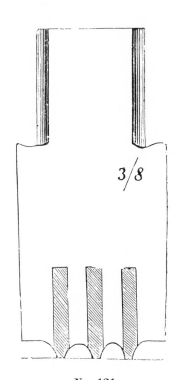

No. 126.　　　　　　　　　　　　　　No. 131.

No.		Price, Each
126.	Center Beads, Double Boxed, . $\frac{1}{8}$, $\frac{3}{16}$, $\frac{1}{4}$, $\frac{5}{16}$, $\frac{3}{8}$, $\frac{7}{16}$, $\frac{1}{2}$ inch,	$0.60
	Center Beads, Double Boxed, ... $\frac{5}{8}$, $\frac{3}{4}$ inch,	.65
	Center Beads, Double Boxed, ... $\frac{7}{8}$, 1 inch,	.80
127.	Center Beads, Solid Boxed, Dovetailed, $\frac{1}{8}$, $\frac{3}{16}$, $\frac{1}{4}$, $\frac{5}{16}$, $\frac{3}{8}$ inch,	.75
	Center Beads, Solid Boxed, Dovetailed, . $\frac{7}{16}$, $\frac{1}{2}$, $\frac{5}{8}$, $\frac{3}{4}$ inch,	.90
131.	Reeding Planes, ... $\frac{1}{4}$, $\frac{5}{16}$, $\frac{3}{8}$, $\frac{1}{2}$ inch,	1.00
128.	Reeding Planes, Cutting two Reeds, . . $\frac{1}{4}$, $\frac{5}{16}$, $\frac{3}{8}$, $\frac{1}{2}$ inch,	1.40
129.	Reeding Planes, Cutting three Reeds . $\frac{1}{4}$, $\frac{5}{16}$, $\frac{3}{8}$, $\frac{1}{2}$ inch,	1.80
130.	Reeding Planes, Cutting four Reeds . $\frac{1}{4}$, $\frac{5}{16}$, $\frac{3}{8}$, $\frac{1}{2}$ inch,	2.30
130$\frac{1}{2}$.	Reeding Planes, Cutting five Reeds . $\frac{1}{4}$, $\frac{5}{16}$, $\frac{3}{8}$, $\frac{1}{2}$ inch,	3.00

NOSING STEP OR STAIR PLANES.

No. 133.

No.   Price, Each

132.	Nosing or Step Plane, one Iron,	.	½, ¾, ⅞, 1, 1⅛, 1¼ inch,	$0.80
	Nosing or Step Plane, one Iron,	. . . . .	1⅜, 1½ inch,	1.00
	Nosing or Step Plane, one Iron,	. . . . .	1¾, 2 inch,	1.25
133.	Nosing or Step Plane, two Irons,	.	½, ¾, ⅞, 1, 1⅛, 1¼ inch,	1.10
	Nosing or Step Plane, two Irons,	. . . .	1⅜, 1½ inch,	1.20
	Nosing or Step Plane, two Irons,	. . . . .	1¾, 2 inch,	1.50
133½.	Nosing or Step Plane, two Irons, Handled,	¾, ⅞, 1, 1⅛, 1¼ inch,		1.50
	Nosing or Step Plane, two Irons, Handled,	. .	1⅜, 1½ inch,	1.75
	Nosing or Step Plane, two Irons, Handled,	. .	1¾, 2 inch,	2.00
134.	Nosing or Step Plane, Single Iron, Handled,	.	1, 1⅛, 1¼ inch,	1 ¦0
	Nosing or Step Plane, Single Iron, Handled,	. .	1⅜, 1½ inch,	1.20
	Nosing or Step Plane, Single Iron, Handled,	. .	1¾, 2 inch,	1.50
135.	Hand Rail Plane, Ovolo or Ogee Handled,	. . . .		1.00

DADOES.

No. 139.

138.	Dado, with Brass Side Stop,	.	¼, $\frac{5}{16}$, ⅜, ½, ⅝, ¾, ⅞, 1 inch,	$1.10
139.	Dado, with Screw Stop,	. .	¼, $\frac{5}{16}$, ⅜, ½, ⅝, ¾, ⅞, 1 inch,	1.50
	Dado, with Solid Handle, add to List Price,	. . . . . .		1.00

DOOR PLANES.

144.   Door Planes, Ogee or Bevel, ...   ½ to ⅝ inch, $0.88

145.   Door Planes, Double Screw Arms, ... 1.50

FILLETSTERS.

No. 148.   **No. 150.**

146.   Filletster, ... $1.10

147.   Filletster, with Stop, ... 1.25

148.   Filletster, with Stop and Cut, ... 1.40

149.   Filletster, with Stop and Cut and Dovetailed, Boxed,   . . . 1.85

150.   Filletster, with Screw Stop, Cut and Dovetailed, Boxed,   . . 2.50

150½.   Filletster, with Screw Stop, Cut and Dovetailed, Boxed and
        Boxwood Fence, ... 3.00

150¾.   Filletster, with Screw Stop, Cut, Solid Box or Rosewood,   . . 5.00

151.   Filletster, with Screw Stop, Cut and Dovetailed, Boxed, Solid
        Handle, ... 3.75

        Extra for Rabbet Mouth, ...25

152.   Filletster, with Arms, Stop, Cut and Dovetailed, Boxed,   . . 2.75

153.   Filletster, with Screw Arms, Stop, Cut and Dovetailed, Boxed,   3.00

154.   Filletster, with Screw Arms, Screw Stop, Cut and Dovetailed,
        Boxed, ... 4.00

154½.   Filletster, with Screw Arms, Screw Stop, Cut Solid Box or
        Rosewood, ... 6.25

154¾.   Back Filletster, Screw Arm and Screw Stop, ... 3.50

RABBET PLANES.

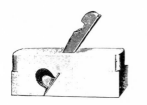

No. 155.

No. 157.

No. 160.

One Pair. No. 161.

No.			Price, Each
155.	Rabbet Planes, Square, ...	¼, ⅜ inch,	$1.00
	Rabbet Planes, Square, ...	½, ⅝, ¾, ⅞, 1 inch,	.60
	Rabbet Planes, Square, ...	1⅛, 1¼ inch,	.65
	Rabbet Planes, Square, ...	1⅜, 1½ inch,	.70
	Rabbet Planes, Square, ...	1¾ inch,	.80
	Rabbet Planes, Square, ...	2 inch,	.90
	Rabbet Planes, Square, ...	2¼ inch,	1.00
156.	Rabbet Planes, Square, Double Boxed,	½, ⅝, ¾, ⅞, 1 inch,	1.05
	Rabbet Planes, Square, Double Boxed, . .	1⅛, 1¼ inch,	1.10
	Rabbet Planes, Square, Double Boxed, . .	1⅜, 1½ inch,	1.15
157.	Rabbet Planes, Skew, ...	¼, ⅜ inch,	1.00
	Rabbet Planes, Skew, ...	½, ⅝, ¾, ⅞, 1 inch,	.60
	Rabbet Planes, Skew, ...	1⅛, 1¼ inch,	.65
	Rabbet Planes, Skew, ...	1⅜, 1½ inch,	.70
	Rabbet Planes, Skew, ...	1⅝, 1¾ inch,	.80
	Rabbet Planes, Skew, ...	1⅞, 2 inch,	.90
	Rabbet Planes, Skew, ...	2¼ inch,	1.00

Extra for One Cutter, add $0.15; for Two Cutters, add $0.30 List.

158.	Rabbet Planes, Skew, Boxed and Cut, .	½, ⅝, ¾, ⅞, 1 inch,	1.05
	Rabbet Planes, Skew, Boxed and Cut, . . .	1⅛, 1¼ inch,	1.10
	Rabbet Planes, Skew, Boxed and Cut, . . .	1⅜, 1½ inch,	1.15
	Rabbet Planes, Skew, Boxed and Cut, . . .	1⅝, 1¾ inch,	1.25
	Rabbet Planes, Skew, Boxed and Cut, ...	1⅞, 2 inch,	1.35
	Rabbet Planes, Skew, Boxed and Cut, ...	2¼ inch,	1.45
159.	Handle Rabbet, 16-in., Handle on Side, one Cut, 1½ to 2 inch,		1.70
	Handle Rabbet, 16-in., Handle on Side, one Cut, 2¼ to 2½ inch,		2.00
160.	Handle Rabbet, 16-in., Handle on Top, one Cut, 1½ to 2 inch,		1.50
	Handle Rabbet, 16-in., Handle on Top, one Cut, 2¼ to 2½ inch,		1.80
	Extra for two Cuts, ...		.15
161.	Side Rabbet Planes, ... per pair,		2.25

RAISING PLANES.

No.		Per Inch
162.	Raising Plane, with Stop and Cut, . . . 2½, 3, 3½, 4 inch,	$1.00
162½.	Raising Plane, Double Iron, with Stop and Cut, 2½, 3, 3½, 4 inch,	1.25
	Shoulder Boxed, ... add to List Price,	.50

HOLLOWS AND ROUNDS.

		Price
163.	Hollows and Rounds, Set 9 pairs, 2 to 18, Even numbers, . .	$9.45
164.	Hollows and Rounds, Set 10 pairs, 2 to 20, Even numbers, . .	10.80
165.	Hollows and Rounds, Set 12 pairs, 2 to 24, Even numbers, . .	13.50
167.	Hollows and Rounds, Nos. 2, 4, 6, 8, 10, 12, . . . per pair,	1.00
	Hollows and Rounds, Nos. 14, 16, 18, ... per pair,	1.15
	Hollows and Rounds, Nos. 20, 22, 24, ... per pair,	1.35
	Hollows and Rounds, Nos. 26, 28, 30, ... per pair,	1.75

HOLLOWS AND ROUNDS, SKEW IRONS.

166.	Hollows and Rounds, Set 9 pairs, 2 to 18, Even numbers, Skewed Irons, ...	11.75
166½.	Hollows and Rounds, 2, 4, 6, 8, 10, 12, Skewed Irons, per pair,	1.20
	Hollows and Rounds, 14, 16, 18, Skewed Irons, . . per pair,	1.55
	Hollows and Rounds, 20, 22, 24, Skewed Irons, . . per pair,	1.75
	Hollows and Rounds, 26, 28, 30, Skewed Irons, . . per pair,	2.00

SIZES OF CIRCLES WORKED BY HOLLOWS AND ROUNDS.

Size No. . . .	2	4	6	8	10	12	14	16	18	20	22	24	26	28	30
Width of Iron, inches,	¼	⅜	½	⅝	¾	⅞	1	1⅛	1¼	1⅜	1½	1⅝	1¾	1⅞	2
Size of Circle worked, inches,	½	¾	1	1¼	1½	1¾	2	2¼	2½	2¾	3	3¼	3½	3¾	4

TABLE HOLLOWS AND ROUNDS.

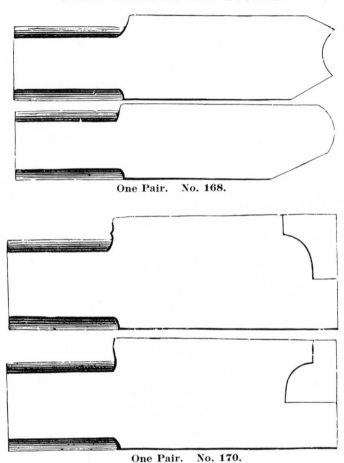

One Pair.   No. 168.

One Pair.   No. 170.

No.		Price, Per Pair
168.	Table Hollows and Rounds, with Gauge ½ and ⅝ inch, . .	$1.35
	Table Hollows and Rounds, with Gauge ¾ inch, ...	1.50
169.	Table Hollows and Rounds, Dovetailed, Boxed, with Gauge ½ and ⅝ inch, ...	1.75
	Table Hollows and Rounds, Dovetailed, Boxed, with Gauge ¾ inch, ...	1.90
170.	Table Hollows and Rounds, with Fence, ½ and ⅝ inch, . .	1.65
	Table Hollows and Rounds, with Fence, ¾ inch, ...	1.85

MATCH PLANES.

No. 171. Double.

No. 174.

No. 175.

No.		Price
171.	Match Planes, Double, one Block, ¼ inch, ... each,	$3.50
	Match Planes, Double, one Block, ⅜, ½, ⅝, ¾, ⅞, 1 inch, each,	1.50
172.	Match Planes, Double, one Block, Plated, ¼ inch, . . . each,	4.25
	Match Planes, Double, one Block, Plated, ⅜, ½, ⅝, ¾, ⅞, 1 inch, ... each,	1.75
173.	Match Planes, ¼ inch, ... per pair,	3.50
	Match Planes, ⅜, ½, ⅝, ¾, ⅞, 1 inch, ... per pair,	1.50
	Match Planes, 1¼, 1½ inch, ... per pair,	2.00
174.	Match Planes, Plated, ¼ inch, ... per pair,	4.25
	Match Planes, Plated, ⅜, ½, ⅝, ¾, ⅞, 1 inch, . . per pair,	1.75
	Match Planes, Plated, 1¼, 1½ inch, ... per pair,	2.25
175.	Match Planes, with Handle, ¼ inch, ... per pair,	4.25
	Match Planes, with Handle, ⅜, ½, ⅝, ¾, ⅞, 1 inch, per pair,	2.25
	Match Planes, with Handle, 1¼, 1½ inch, ... per pair,	2.75
176.	Match Planes, with Handle, Plated, ¼ inch, . . . per pair,	4.50
	Match Planes, with Handle, Plated, ⅜, ½, ⅝, ¾, ⅞, 1 inch, ... per pair,	2.50
	Match Planes, with Handle, Plated, 1¼, 1½ inch, . per pair,	3.00

SASH PLANES—Ovolo, Bevel, Gothic and Ogee.

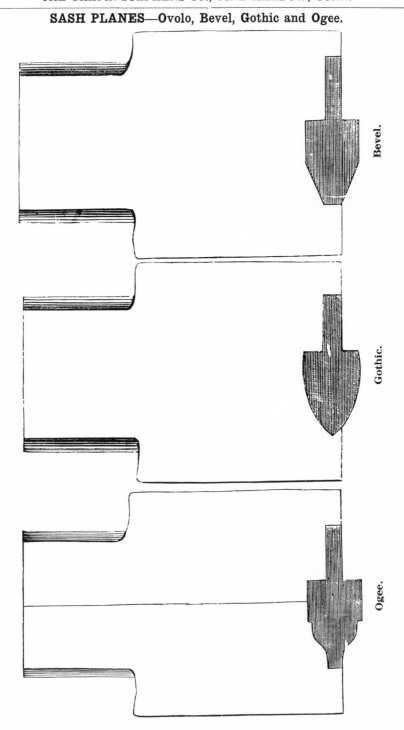

SASH PLANES.

OVOLO, BEVEL, GOTHIC AND OGEE.

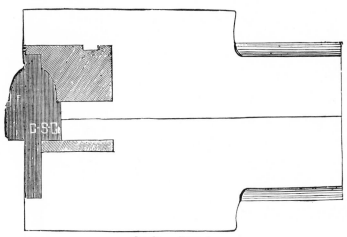

Ovolo. Boxed.

No.		Price, Each
185.	Sash Plane, with one Iron, Bevel or Ovolo, . . .	1½ inch, $0.60
189.	Sash Plane, with two Irons, Bevel or Ovolo, . . .	1½ inch, 1.00
191.	Sash Plane, with two Irons, Boxed, Bevel or Ovolo, .	1½ inch, 1.25
187.	Sash Plane, two Irons, Self-Regulating and Brass Pad, Ovolo, ...	1½ inch, 1.50
188.	Sash Plane, two Irons, Self-Regulating and Brass Pad, Boxed, Ovolo, ...	1½ inch, 1.65
192.	Sash Plane, Screw Arms, Self-Regulating, Bevel or Ovolo, ...	1½ inch, 1.75
192½.	Sash Plane, Screw Arms, Self-Regulating, Gothic or Ogee, ...	1½ inch, 1.75
195.	Sash Plane, Screw Arms, Self-Regulating, Boxed, Bevel or Ovolo, ...	1½ inch, 2.00
195½.	Sash Plane, Screw Arms, Self-Regulating, Boxed, Gothic or Ogee, ...	1½ inch, 2.00
196.	Sash Plane, Screw Arms, Self-Regulating, Dovetailed, Boxed, Bevel or Ovolo, ...	1½ inch, 2.25
196½.	Sash Plane, Screw Arms, Self-Regulating, Dovetailed, Boxed, Gothic or Ogee, ...	1½ inch, 2.25

**Nos. 185, 187, 188, 189, 191, 192, 192½, 195, 195½ will be furnished
1¼ inch size if so desired.**

SASH COPING PLANES.

OVOLO, BEVEL, GOTHIC AND OGEE.

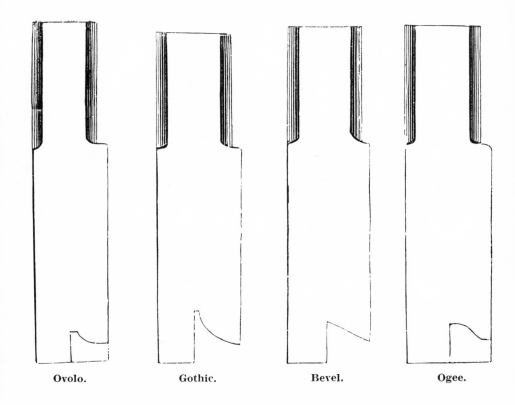

Ovolo.   Gothic.   Bevel.   Ogee.

No.		Price, Each
200.	Sash Coping Planes, Double, ...	$0.75
200½.	Sash Coping Planes, Single, ...	.50
201.	Sash Coping Planes, Double and Boxed, ...	.90

MOULDING PLANES.

Size Given is the Width the Tool Works.   Depths Usually One-Half the Width.

PLAIN OGEES.

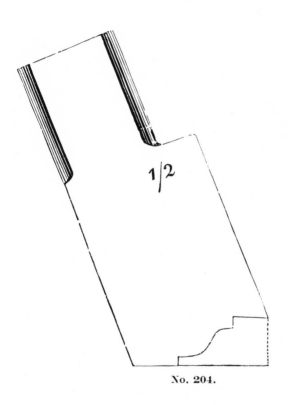

No. 204.

No.		Price, Each
204.	Plain Ogee, ... ⅜, ½, ⅝, ¾, ⅞, 1 inch,	$0.65
	Plain Ogee, ... 1¼, 1½ inch,	.85
	Plain Ogee, ... 1¾, 2 inch,	1.00
205.	Plain Ogee, with Bead, ... ⅜, ½, ⅝, ¾, ⅞, 1 inch,	.75
	Plain Ogee, with Bead, ... 1¼, 1½ inch,	1.00
	Plain Ogee, with Bead, ... 1¾, 2 inch,	1.25

REVERSE OGEES.

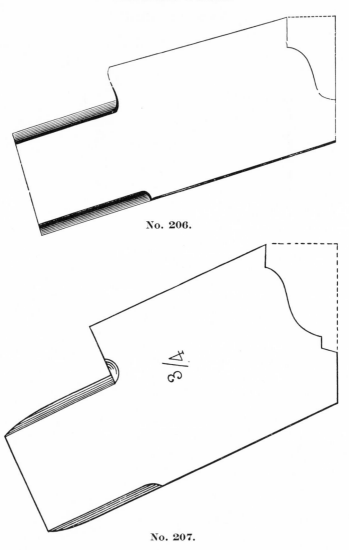

No. 206.

No. 207.

No.		Price, Each
206.	Reverse and Back Ogee, ...	¾, ⅞, 1 inch, $0.75
	Reverse and Back Ogee, ...	1¼, 1½ inch, .85
	Reverse and Back Ogee, ...	1¾, 2 inch, 1.00
	Reverse and Back Ogee, ...	2¼, 2½ inch, 1.25
207.	Reverse and Back Ogee, with Bead or Square,	⅜, ½, ⅝ inch, .75
	Reverse and Back Ogee, with Bead or Square,	¾, ⅞, 1 inch, .85
	Reverse and Back Ogee, with Bead or Square,	1¼ inch, 1.00
	Reverse and Back Ogee, with Bead or Square,	1½, 1¾ inch, 1.15

REVERSE OGEES.

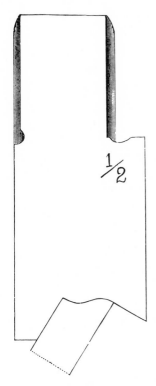

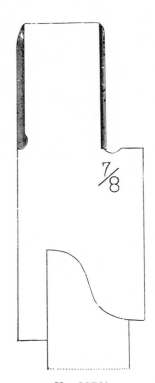

No. 206½.   **No. 207½.**

No.		Price, Each
206½.	Reverse Ogee, with Fence, ...	⅜, ½, ⅝ inch, $0.75
	Reverse Ogee, with Fence, ...	¾, ⅞, 1 inch, .85
	Reverse Ogee, with Fence, ...	1¼ inch, 1.00
	Reverse Ogee, with Fence, ...	1½ inch, 1.15
207½.	Roman Reverse Ogee, with Fence, ...	⅜, ½, ⅝ inch, .80
	Roman Reverse Ogee, with Fence, ...	¾, ⅞, 1 inch, .90
	Roman Reverse Ogee, with Fence, ...	1¼ inch, 1.05
	Roman Reverse Ogee, with Fence, ...	1½ inch, 1.20
	With Handle, extra, add to List Price, ...	.50

GRECIAN OGEES.

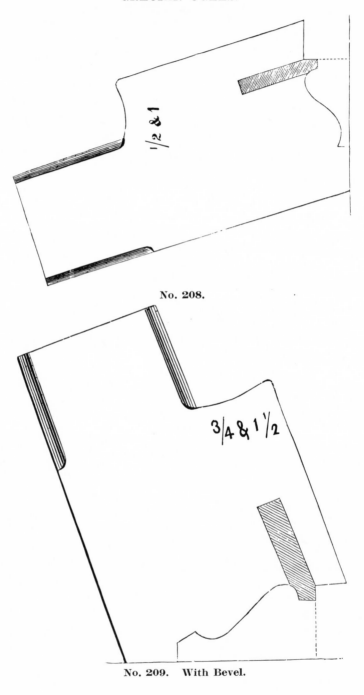

1/2 & 1

No. 208.

3/4 & 1 1/2

No. 209.   With Bevel.

GRECIAN OGEES.

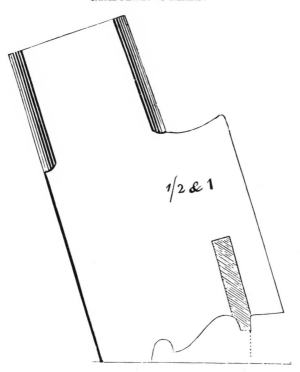

No. 209. With Bead.

No.		Price, Each
208.	Grecian Ogee, ... ⅜, ½, ⅝, ¾, ⅞, 1 inch,	$0.75
	Grecian Ogee, ... 1¼, 1½ inch,	.90
	Grecian Ogee, ... 1¾, 2 inch,	1.10
	Grecian Ogee, with Handle, add to List Price, ...	.50
209.	Grecian Ogee, with Bevel, Fillet, or Bead,	
	⅜, ½, ⅝, ¾, ⅞, 1 inch,	.90
	Grecian Ogee, with Bevel, Fillet, or Bead, . 1¼, 1½ inch,	1.05
	Grecian Ogee, with Bevel, Fillet, or Bead, . 1¾, 2 inch,	1.25
	Grecian Ogee, with Bevel, Fillet, or Bead, with Handle, add to List Price, ...	.50
210.	Quirk Ogees, ... ⅜, ½, ⅝, ¾, ⅞, 1 inch,	.70
	Quirk Ogees, ... 1¼, 1½ inch,	.85
211.	Quirk Ogees, with Bead, ... ⅜, ½, ⅝, ¾, ⅞, 1 inch,	1.25
	Quirk Ogees, with Bead, ... 1¼ inch,	1.40
	Quirk Ogees, with Bead, ... 1½ inch,	1.60
	Quirk Ogees, with Bead, ... 1¾ inch,	1.80
	Quirk Ogees, with Bead, ... 2 inch,	2.00

GRECIAN OVOLOS.

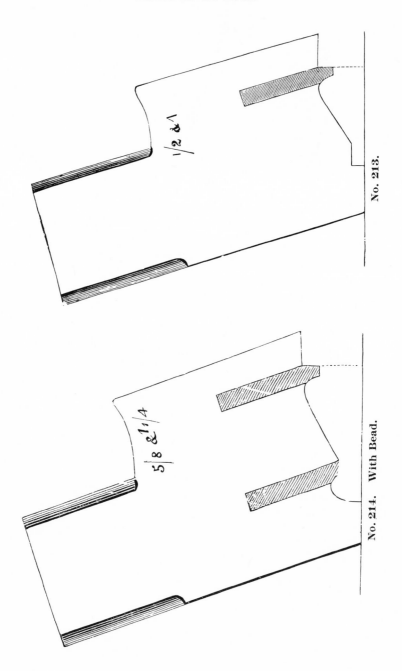

GRECIAN OVOLOS.

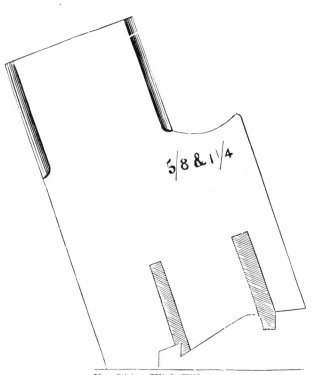

No. 214. With Fillet.

No.		Price, Each
212.	Grecian Ovolo, ... ½, ⅝, ¾, ⅞, 1 inch,	$0.75
	Grecian Ovolo, ... 1¼, 1½ inch,	.90
	Grecian Ovolo, ... 1¾, 2 inch, ·	1.10
	Grecian Ovolo, with Handle, add to List Price, ...	.50
213.	Grecian Ovolo, with Square, ... ½, ⅝, ¾, ⅞, 1 inch,	1.00
	Grecian Ovolo, with Square, ... 1¼, 1½ inch,	1.10
	Grecian Ovolo, with Square, ... 1¾, 2 inch,	1.25
	Grecian Ovolo, with Square, with Handle, add to List Price, .	.50
214.	Grecian Ovolo, with Bead, or Fillet, . ½, ⅝, ¾, ⅞, 1 inch,	1.00
	Grecian Ovolo, with Bead, or Fillet, . . . 1¼, 1½ inch,	1.10
	Grecian Ovolo, with Bead, or Fillet, . . . 1¾, 2 inch,	1.25
	Grecian Ovolo, with Bead, or Fillet, with Handle, add to List Price, ...	.50
215.	Quirk Ovolo, ... ½, ⅝, ¾, ⅞, 1 inch,	.70
	Quirk Ovolo, ... 1¼, 1½ inch,	.80
216.	Quirk Ovolo, with Bead, ... ½, ⅝, ¾, ⅞, 1 inch,	.85
	Quirk Ovolo, with Bead, ... 1¼, 1½ inch,	1.00

QUARTER ROUNDS AND OVOLOS.

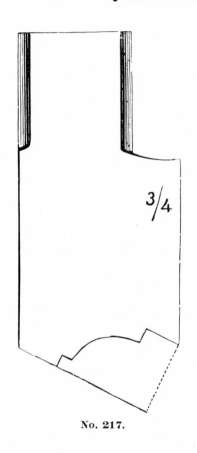

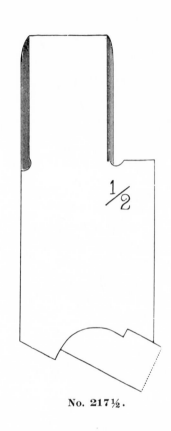

No. 217.   No. 217½.

No.			Price, Each
217.	Quarter Round or Ovolo, two Squares, .	¼, ⅜, ½, ⅝ inch,	$0.50
	Quarter Round or Ovolo, ...	¾, ⅞, 1 inch,	.65
	Quarter Round or Ovolo, ...	1¼, 1½ inch,	.75
218.	Quarter Round, with Bead and one Square,	⅜, ½, ⅝, ¾ inch,	.65
	Quarter Round, with Bead, ...	⅞, 1 inch,	.80
	Quarter Round, with Bead, ...	1¼, 1½ inch,	.90
217½.	Quarter Round or Casing Moulding, one Square,	⅜, ½, ⅝ inch,	.50
	Quarter Round or Casing Moulding, one Square,	¾, ⅞, 1 inch,	.65
	Quarter Round or Casing Moulding, one Square,	1¼, 1½ inch,	.75

QUARTER ROUNDS AND SCOTIAS.

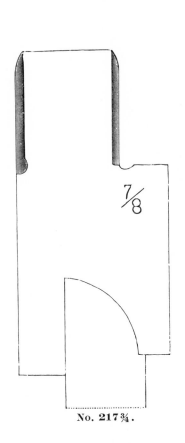

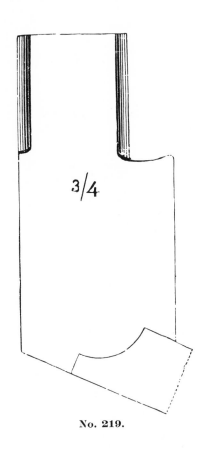

No. 217¾.   No. 219.

No.		Price, Each
217¾.	Quarter Round or Casing Moulding, with Fence, ⅜, ½, ⅝, ¾ inch,	$0.80
	Quarter Round or Casing Moulding, with Fence, . ⅞, 1 inch,	.90
	Quarter Round or Casing Moulding, with Fence, 1¼, 1½ inch,	1.00
	Quarter Round or Casing Moulding, with Handle, add to List Price, ...	.50
219.	Scotia or Cove, ... ⅜, ½, ⅝, ¾ inch,	.50
	Scotia or Cove, ... ⅞, 1 inch,	.65
	Scotia or Cove, ... 1¼, 1½ inch,	.75
220.	Scotia or Cove, with Bead, ... ⅜, ½, ⅝, ¾ inch,	1.25
	Scotia or Cove, with Bead, ... ⅞, 1 inch,	1.50
	Scotia or Cove, with Bead, ... 1¼, 1½ inch,	1.75

SNIPE BILLS AND BASE MOULDINGS.

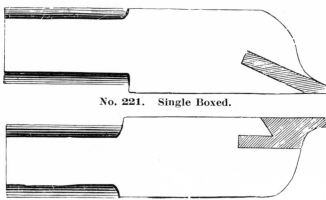

No. 221.   Single Boxed.

No. 221½.   Full Boxed.

No.		Price
221.	Snipe Bills, ...	per pair, $2.25
221½.	Snipe Bills, Full Boxed, ...	per pair, 2.75

7/8 & 3

No. 222.   Base Moulding.

			Price, Each
222.	Base Moulding Plane, with Handle, . . .	2, 2¼, 2½ inch,	$2.50
	Base Moulding Plane, with Handle, ...	2¾, 3 inch,	3.00
223.	Bed Moulding Plane, with Handle, . . .	2, 2¼, 2½ inch,	2.50
	Bed Moulding Plane, with Handle, ...	2¾, 3 inch,	3.00

CORNICE, CABINET AND HALVING PLANES.

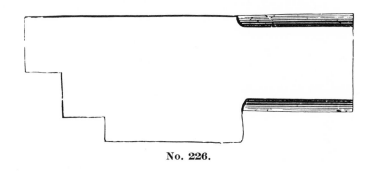

No. 226.

No. 224.

PAIR CORNICE PLANES.

No.		Price
224.	Cornice Plane, ... per inch,	$1.00
225.	Cabinet Ogee Planes, ... pcr inch,	1.00
226.	Halving Plane, ... each,	.75
227.	Halving Plane, Handle and Plated, ... each,	1.40

Moulding Planes made from Samples or Drawings on Short Notice.

GROVING PLOWS.

No. 232.

No. 234.

Applewood Screw Arms, Single Plate.

No.		Price, Each
230.	Plow, Wood Stop, four Irons, ...	$4.00
231.	Plow, Wood Stop, Boxed or Plated Fence, four Irons, . . .	4.25
232.	Plow, Screw Stop, eight Irons, ...	5.50
233.	Plow, Screw Stop, Boxed or Plated Fence, eight Irons, . . .	5.85
234.	Plow, Screw Stop, Solid Handle, eight Irons, ...	6.50
235.	Plow, Screw Stop, Solid Handle, Boxed or Plated Fence, eight Irons, ...	6.85

GROOVING PLOWS.

No. 238.

Solid Handle, Boxwood Screw Arms, and Screw Stop with Side Stop.

No.		Price, Each
236.	Handled Plow, Beech, Single Plate, eight Irons, ...	$7.75
237.	Handled Plow, Beech, Boxed Fence, Single Plate, eight Irons,	7.85
238.	Handled Plow, Beech, Boxed Fence, Best Plate, eight Irons, .	8.00
239.	Handled Plow, Beech, Boxed Fence, Best Plate, Polished, eight Irons, ...	8.50
239½.	Handled Plow, Applewood, Boxed Fence, Best Plate, Polished, eight Irons, ...	9.25
240.	Handled Plow, Solid Boxwood, Best Plate, Polished, eight Irons,	10.00
240½.	Handled Plow, Solid Rosewood, Boxed Fence, Best Plate, Polished, eight Irons, ...	10.25

GROOVING PLOWS.

No. 244.

Boxwood Screw Arms, and Screw Stop, with Side Stop.

No.		Price, Each
242.	Plow, Single Plate, eight Irons, ...	$6.25
243.	Plow, Boxed Fence, Single Plate, eight Irons, ...	6.50
244.	Plow, Boxed Fence, Best Plate, eight Irons, ...	6.60
244½.	Plow, Applewood, Best Plate, Polished, eight Irons, . . .	7.50
245.	Plow, Solid Boxwood, Best Plate, Polished, eight Irons, . .	8.00
245½.	Plow, Solid Rosewood, Boxed Fence, Best Plate, Polished, eight Irons, ...	8.25
	Plows, with Skate Iron Pattern Plates, extra, ...	1.00
	Plows with Steel Facing on Fence, extra, ...	.50
	Boxwood Plow Arms, complete, per pair, . . . net,	$1.75
	Applewood Plow Arms, complete, per pair, . . net,	1.00

MARKING GAUGES.

One Dozen in a Box.

No. 246.

No.		Per Dozen
246.	Common Marking Gauge, with inches, ...	$0.75
247.	Common Marking Gauge, Oval Bar, with inches, ...	1.00

PREMIUM GAUGES.

One Dozen in a Box.

No. 248.

248.	Marking Gauge, Oval Head and Bar, Steel Points, with inches,	$1.25
271.	Marking Gauge, Polished, Oval Bar, Steel Points, with inches,	2.00
272.	Marking Gauge, Polished, Plated Head, Oval Bar, Steel Points,	
	with inches, ...	2.75
249.	Marking Gauge, Applewood, Oval Head and Bar, Steel Points,	
	with inches, ...	2.00
250.	Marking Gauge, Mahogany or Applewood, Plated Oval Head	
	and Bar, Steel Points, with inches, ...	4.00

PREMIUM GAUGES.

One Dozen in a Box.

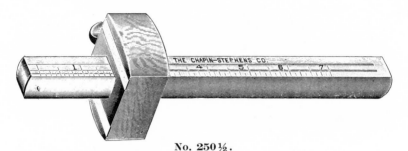

No. 250½.

No.   Per Dozen

250½. Marking Gauge, Mahogany or Applewood, Plated Head and
    Bar, Brass Thumb Screw, Oval Bar, Steel Points, with
    inches, ... $5.50

251. Marking Gauge, Box or Rosewood, Oval Head and Bar, Brass
    Thumb Screw, Steel Points, with inches, ... 3.50

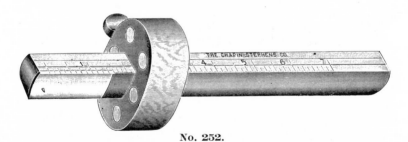

No. 252.

252. Marking Gauge, Box or Rosewood, Plated Oval Head and Bar,
    Brass Thumb Screw, Steel Points, with inches, ... 6.00

CUTTING GAUGES.

One Dozen in a Box.

253. Cutting Gauge, Oval Bar, Steel Cutters, with inches, . . . 2.50

254. Cutting Gauge, Applewood, Oval Bar, Steel Cutters, with inches, 3.25

PANEL GAUGES.

One-Half Dozen in a Box.

No.		Per Dozen
255.	Panel Gauge, Oval Bar, Brass Thumb Screw, Steel Points, with inches, ...	$4.75
256.	Panel Gauge, Applewood, Oval Bar, Brass Thumb Screw, Steel Points, with inches, ...	6.00
257.	Panel Gauge, Mahogany, Plated Head and Bar, Oval Bar, Brass Thumb Screw, Steel Points, with inches, ...	12.00
258.	Panel Gauge, Rosewood, Plated Head and Bar, Oval Bar, Brass Thumb Screw, Steel Points, with inches, ...	18.00

MORTISE GAUGES.

One-Half Dozen in a Box.

259.	Mortise Gauge, Mahogany or Applewood, Plated Head, Thumb Slide, Brass Thumb Screw, Steel Points, ...	6.50
259½.	Mortise Gauge, Mahogany or Applewood, Plated Head and Bar, Thumb Slide, Brass Thumb Screw, Steel Points, . . .	8.00

No. 260.

260.	Mortise Gauge, Mahogany or Applewood, Plated Head, Screw Slide, Brass Thumb Screw, Steel Points, ...	9.00
261.	Mortise Gauge, Box or Rosewood, Plated Head, Screw Slide, Brass Thumb Screw, Steel Points, ...	11.00
262.	Mortise Gauge, Box or Rosewood, Plated Head and Bar, Screw Slide, Brass Thumb Screw, Steel Points, ...	14.00
263.	Mortise Gauge, Box or Rosewood, Full Plated Head, Screw Slide, Brass Thumb Screw, Steel Points, ...	18.00
264.	Mortise Gauge, Box or Rosewood, Full Plated Head and Bar, Screw Slide, Brass Thumb Screw, Steel Points, ...	20.00

LEVELS.

Hardwood, 12 Inch.

No. 100.

With Handsome Nickel Top Plate.

Packed One Dozen in a Pasteboard Box.

No.		Per Dozen
100.	Oil finish, ...	$1.25
100.	Polished, ...	1.40

PLUMBS AND LEVELS.

Hardwood, 12 Inch.

No. 102.

With Handsome Nickel Top Plate.

Packed One Dozen in a Pasteboard Box.

102.	Oil finish, ...	$2.20
102.	Polished, ...	2.50

In ordering always specify whether oil finish or polished Plumbs and Levels are wanted, otherwise polished goods will be shipped.

LEVELS.

Highly Hand Polished.

No. 286.

No.		Per Dozen
286.	Cherry, Polished, Nickel Arch Top Plate, Side Views, Assorted, 10 to 16-inch, ...	$4.25
287.	Cherry, Polished, Nickel Arch Top Plate, Side Views, Assorted, 18 to 24-inch, ...	5.35

PLUMBS AND LEVELS.

No. 289.

289.	Cherry, Polished, Nickel Arch Top Plate, Side Views, Assorted, 12 to 18-inch, ...	6.35
289½.	Cherry, Polished, Nickel Arch Top Plate, Side Views, Assorted, 18 to 24-inch, ...	7.30

No. 290.

290.	Cherry, Polished, Nickel Arch Top Plate, Side Views, Assorted, 24 to 30-inch, ...	8.25
290½.	Cherry, Polished, Brass Arch Top Plate, Brass Lipped Side Views, Assorted, 24 to 30-inch, ...	11.00
291.	Mahogany, Polished, Brass Arch Top Plate, Side Views, Assorted, 24 to 30-inch, ...	12.25
292.	Mahogany, Polished, Brass Arch Top Plate, Brass Lipped Side Views, Assorted, 24 to 30-inch, ...	17.75

PLUMBS AND LEVELS.

No. 293.

No.   Per Dozen

293.   Cherry, Polished, Brass Arch Top Plate, Brass-Lipped Side
         Views and Tipped, Assorted, 24 to 30-inch, ... $15.75
293½.   Cherry, Polished, Brass Arch Top Plate, Side Views and
         Tipped, Assorted, 24 to 30-inch, ...   12.25
294.   Cherry, Triple Stock, Brass Arch Top Plate, Polished, Brass-
         Lipped Side Views and Tipped, Assorted, 24 to 30-inch,   19.00
295.   Mahogany, Polished, Brass Arch Top Plate, Side Views, As-
         sorted, 18 to 24-inch, ...   11.60

No. 296.

296.   Mahogany, Polished, Brass Arch Top Plate, Brass-Lipped Side
         Views and Tipped, Assorted, 12 to 18-inch, ...   12.25
297.   Mahogany, Polished, Brass Arch Top Plate, Brass-Lipped Side
         Views and Tipped, Assorted, 24 to 30-inch, ...   23.00

No. 300.

300.   Rosewood, Polished, Brass Arch Top Plate, Brass-Lipped Side
         Views and Tipped, 24 to 30-inch, ...   40.00

Note.—Nos. 290 and 293½ we also keep in stock assorted, 12 to 22 inches. The
regular assortment, 24 to 30 inches, will always be sent unless otherwise ordered.

PATENT ADJUSTABLE PLUMBS AND LEVELS.

No. 490½.

No. Per Dozen

490½.   Cherry, Patent Adjustable, Brass Arch Top Plate, Brass-Lipped
Side Views, Polished, Assorted, 24 to 30-inch, ... $12.25

No. 493.

493.   Cherry, Patent Adjustable, Brass Arch Top Plate, Brass-Lipped
Side Views and Tipped, Polished, Assorted, 24 to 30-inch,   16.00

No. 493½.

493½.   Cherry, Patent Adjustable, Brass Arch Top Plate, Side Views
and Tipped, Polished, Assorted, 24 to 30 inch,   . . .   13.75

No. 494.

494.   Cherry, Patent Adjustable, Brass Arch Top Plate, Triple Stock,
Brass-Lipped Side Views and Tipped, Pol., Asst'd, 24 to 30-in.,   20.50

491.   Mahogany, Patent Adjustable, Brass Arch Top Plate, Side Views,
Polished, Assorted, 24 to 30-inch, ...   16.50

492.   Mahogany, Patent Adjustable, Brass Arch Top Plate, Brass-
Lipped Side Views, Polished, 24 to 30-inch, ...   19.75

497.   Mahogany, Patent Adjustable, Brass Arch Top Plate, Brass-
Lipped Side Views and Tipped, Pol., Assorted, 24 to 30-inch,   23.00

Note.—No. 493½ we also keep in stock, assorted, 12 to 22 inches.   The regular
assortment, 24 to 30 inches, will always be sent unless otherwise ordered.

PLUMBS AND LEVELS,

WITH GRADUATING ADJUSTMENTS,

To Work at any Angle or Elevation Required.

No. 304.

No.		Per Dozen
304.	Cherry, Graduated Adjusting, Brass Arch Top Plate, Polished, with Side Views, Assorted, 24 to 30-inch, ...	$14.00
306.	Cherry, Graduated Adjusting, Brass Arch Top Plate, Polished, Brass-Lipped Side Views and Tipped, 24 to 30-inch, . .	20.00

PATENT ADJUSTABLE PLUMBS AND LEVELS,

WITH GRADUATING ADJUSTMENTS.

To Work at any Angle or Elevation Required.

504.	Cherry, Patent Adjustable, with Graduating Adjustment, Brass Arch Top Plate, Brass-Lipped Side Views, Polished, Assorted, 24 to 30-inch, ...	17.00
506.	Cherry, Patent Adjustable, with Graduating Adjustment, Brass Arch Top Plate, Brass-Lipped Side Views and Tipped, Polished, Assorted, 24 to 30-inch, ...	23.00

MASONS' PLUMBS AND LEVELS.

310.	Cherry, Polished, Brass Arch Top Plate, Two Plumbs, Side Views and Tipped, 36-inch, ...	18.00
310¼.	Cherry, Polished, Brass Arch Top Plate, Two Plumbs, Side Views, 36-inch, ...	15.75
310½.	Cherry, Polished, Arch Top Plate, Two Plumbs, Side Views, 42-inch, ...	18.00

PATENT ADJUSTABLE MASONS' PLUMBS AND LEVELS.

410¼.	Cherry, Polished, Brass Arch Top Plate, Patent Adjustable, Two Plumbs, Side Views, 36-inch, ...	19.00
490¾.	Cherry, Polished, Brass Arch Top Plate, Patent Adjustable, Two Plumbs, Brass-Lipped Side Views, 42-inch, . . .	22.00
35.	Cherry, Patent Adjustable, for Plumb Line, 42-inch, . . .	19.75

PLUMBS AND LEVELS.

CHAPIN'S IMPROVED BRASS CORNERED.

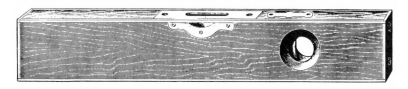

Patent Adjustable Plumbs and Levels, Brass Corners, Heavy Brass Top Plates and Ends, Brass-Lipped Side Views.

End view of our Brass Cornered Level, with tip removed, showing Solid Brass Corners Dovetailed into Wood.

No. 1. SOLID CHERRY.
Extra Quality.

Length,	18	24	26	28	30 inch
Price, Each,	$3.50	$4.00	$4.25	$4.50	$5.00

No. 2. SOLID MAHOGANY.
Extra Quality.

Length,	18	24	26	28	30 inch
Price, Each,	$4.00	$4.85	$5.00	$5.15	$5.50

No. 3. SOLID ROSEWOOD.
Extra Quality.

Length,	18	24	26	28	30 inch
Price, Each,	$5.50	$7.00	$7.25	$7.60	$8.00

MACHINISTS' LEVELS.

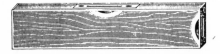

No. 10.

Rosewood Plumb and Level, Brass Corners, Heavy Brass Top Plates and Ends, Brass Side Views.

No. 10.   SOLID ROSEWOOD.

Extra Quality.

Length,	8	10	12 inch
Price, Each,	$2.00	$2.40	$2.60

No. 12.

12.   Solid Rosewood Level (no Plumb), Extra Quality, Brass Corners, Heavy Brass Top Plate and Ends, Brass Side Views. Length, 6½ inches, Price, Each, ... $1.10

All brass cornered Plumbs and Levels are highly polished the natural color of the wood.

Prices on Masons' Plumb Rules and Plumb Rules with Level, quoted on application.

EXTENSION SIGHT LEVELS.

WOOD'S PATENT.

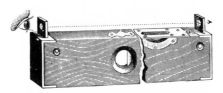

Sights Raised for Long Distance.

Sights Depressed for Ordinary Use.

No.		Per Dozen
1½A.	High Polished Cherry, Brass Lipped, Brass Tipped, Assorted, 18 to 28, ½ doz. in a case, 1-18, 1-24, 2-26, 2-28, . . .	$33.00
1½B.	High Polished Cherry, Brass Lipped, Brass Tipped, Assorted, 24 to 30, ½ doz. in a case, 1-24, 2-26, 2-28, 1-30, . . .	33.00

This improvement greatly enlarges the scope of the tool over the usual form of level, making it available for leveling lengths of **100 feet or more**, and shorter lengths, without the cumbersome and unreliable straight edge commonly used with the ordinary level.

This is accomplished by providing sight pieces working automatically at the ends of the level, whereby the line of sight may be extended indefinitely.

When not in use these sights are depressed flush with the surface of the level. By pressing the buttons in the side of the level the two sights are thrown into position.

These sights are accurately adjusted and may be relied upon to extend the horizontal line to any distance required with accuracy.

These Levels with the extension sights serve a purpose which **NO OTHERS CAN ACCOMPLISH.**

POCKET LEVELS.

No.		Per Dozen
311.	Iron Pocket Level with Nickel Top Plate, ...	$1.30
312.	Iron Pocket Level with Brass Top Plate, ...	1.40

LEVEL GLASSES.

PLAIN.

Length,	1 in.	1¼ in.	1½ in.	1¾ in.	2 in.	2¼ in.
List, Per Gross,	$9.48	$9.60	$9.72	$9.84	$10.00	$10.25
Length,	2½ in.	3 in.	3½ in.	4 in.	4½ in.	Asst'd.
List, Per Gross,	$10.50	$11.50	$13.00	$14.50	$16.00	$12.00

MARKED WITH TWO SILVER LINES.

Length,	1 in.	1¼ in.	1½ in.	1¾ in.	2 in.	2¼ in.
List, Per Gross,	$9.48	$9.60	$9.72	$9.84	$10.00	$10.25
Length,	2½ in.	3 in.	3½ in.	4 in.	4½ in.	Asst'd.
List, Per Gross,	$10.50	$11.50	$13.00	$14.50	$16.00	$12.00

HAND SCREWS, WITH HICKORY SCREWS.

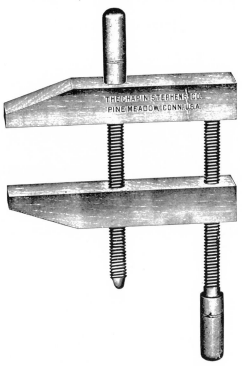

No.	Diam. of Screw	Lgth. of Screw	Lgth. of Jaw	Size of Jaw	To Open	List Price, Per Dozen
800.	1¼ inch,	28 inch,	24 inch,	3 x 3 inch,	17 inch,	$40.00
801.	1¼ inch,	26 inch,	22 inch,	2¾ x 2¾ inch,	15½ inch,	35.00
802.	1¼ inch,	24 inch,	20 inch,	2⅝ x 2⅝ inch,	13¾ inch,	32.00
803.	1¼ inch,	22 inch,	20 inch,	2½ x 2½ inch,	12 inch,	30.00
804.	1⅛ inch,	22 inch,	18 inch,	2½ x 2½ inch,	12¼ inch,	28.50
805.	1⅛ inch,	20 inch,	18 inch,	2⅜ x 2⅜ inch,	10½ inch,	27.00
806.	1 inch,	20 inch,	16 inch,	2⅜ x 2⅜ inch,	11 inch,	25.00
807.	1 inch,	18 inch,	16 inch,	2¼ x 2¼ inch,	9¼ inch,	23.50
808.	⅞ inch,	18 inch,	14 inch,	2⅛ x 2⅛ inch,	10 inch,	22.00
809.	⅞ inch,	16 inch,	14 inch,	2 x 2 inch,	8¼ inch,	20.00
810.	⅞ inch,	16 inch,	12 inch,	1⅞ x 1⅞ inch,	8½ inch,	18.50
811.	¾ inch,	14 inch,	12 inch,	1¾ x 1¾ inch,	7¼ inch,	17.00
812.	¾ inch,	12 inch,	10 inch,	1⅝ x 1⅝ inch,	5½ inch,	14.50
813.	⅝ inch,	10 inch,	8 inch,	1⅜ x 1⅜ inch,	4½ inch,	12.00
814.	⅝ inch,	8 inch,	7 inch,	1⅛ x 1⅛ inch,	3 inch,	9.50
815.	½ inch,	6 inch,	5 inch,	1 x 1 inch,	2 inch,	8.00
816.	⅜ inch,	5 inch,	4 inch,	⅞ x ⅞ inch,	1¼ inch,	7.00
817.	Jewelers' Thumb Screws, ⅜ inch,	. . . .			. . .	7.00

Extra Quality, with Beaded Jaws, 25 Cents per Dozen, Net, Extra.
Either Screw or Jaw one-third the price of the complete Hand Screw.
Moulders' Flask Screws and Press Screws made to order.

CHISEL HANDLES.

One Dozen in a Box.

WITH POLISHED BRASS TUBING FERRULES.

No.		Per Gross
361.	Hickory, four sizes, assorted, 1 to 2 inch, ...	$10.00
362.	Hickory, six sizes, assorted, 1/4 to 2 inch, ...	9.50
363.	Hickory, four sizes, assorted, 1/8 to 1 1/2 inch, ...	8.75
364.	Applewood, four sizes, assorted, 1 to 2 inch, ...	11.00
365.	Applewood, six sizes, assorted, 1/4 to 2 inch, ...	10.50
366.	Applewood, four sizes, 1/8 to 1 1/2 inch, ...	9.75
367.	Rosewood, or Turkey Boxwood, extra finish, 1/8 to 2 inch, . .	24.00

WITH SEAMLESS BRASS FERRULES.

368.	Hickory, four sizes assorted, 1 to 2 inch, ...	7.50
369.	Hickory, six sizes, assorted, 1/4 to 2 inch, ...	6.75
370.	Hickory, four sizes, assorted, 1/8 to 1 1/2 inch, ...	6.00
371.	Applewood, four sizes, assorted, 1 to 2 inch, ...	8.50
372.	Applewood, six sizes, assorted, 1/4 to 2 inch, ...	7.50
373.	Applewood, four sizes, assorted, 1/8 to 1 1/2 inch, ...	6.50

Nos. 368 to 373 inclusive, with Polished Seamless Iron Ferrules, extra, $1.00.

SOCKET CHISEL HANDLES.

POLISHED.

No.		Per Gross
374.	Socket Firmer, Hickory, assorted, ...	$4.50
375.	Socket Firmer, Applewood, assorted, ...	5.50

376.	Socket Framing, Hickory, assorted, ...	5.00

377.	Socket Framing, Hickory, Iron Ferrule, assorted, ...	8.00

FILE HANDLES.

Three Dozen in a Box.

SEAMLESS BRASS FERRULES.

378.	Cherry, assorted, four sizes, ...	5.00
379.	Cherry, assorted, three large sizes, ...	5.50
380.	Soft Wood, assorted, four sizes, ...	4.00
381.	Soft Wood, assorted, three large sizes, ...	4.50
382.	Rosewood, or Turkey Boxwood, extra finish, Jewelers' use, .	18.00
	File Handles with Polished Seamless Iron Ferrules, extra, . .	1.00

BRAD-AWL HANDLES.

Three Dozen in a Box.

SEAMLESS BRASS FERRULES.

No.		Per Gross
383.	Cherry assorted, four sizes, ...	$4.00
384.	Cherry, assorted, three large sizes, ...	4.50
385.	Applewood, assorted, four sizes, ...	5.00
386.	Applewood, assorted, three large sizes, ...	5.50
387.	Rosewood, or Turkey Boxwood, extra finish, ...	18.00

AWL HAFTS.

Three Dozen in a Box.

388.	Sewing Awl Hafts, Brass Ferrules, ...	5.00
389.	Sewing Awl Hafts, Applewood, Brass Ferrules, ...	6.00

390.	Pegging Awl Hafts, Hickory, Brass Ferrules, ...	4.25

SCRATCH AWL HANDLES.

Three Dozen in a Box.

No.		Per Gross
391.	Cherry, ...	$5.50
392.	Applewood, ...	6.50
393.	Rosewood, ...	12.00

SCREW DRIVER HANDLES.

Three Dozen in a Box.

394.	Applewood, Flat, Slotted Ferrule, ...	20.00
395.	Round, Slotted Ferrule, ...	10.00

CARVING TOOL HANDLES.

396.	Beech, four sizes, ½ to ¾ inch, ...	6.00
397.	Boxwood, four sizes, ½ to ¾ inch, ...	20.00
398.	Rosewood, four sizes, ½ to ¾ inch, ...	24.00

PLANE HANDLES.

MADE FROM SELECTED BEECH.

Packed in Paper Boxes, One Dozen in a Box.

No. 1.   **No. 1½.**   **No. 3.**

No.		Per Dozen
1.	Jack Plane Handle, for Wood Planes, ...	$0.50
1½.	Jack Plane Handles, for Iron Planes, ...	.55
2.	Jack Plane Handles, with Bolts, for Wood Planes, ...	1.25
2½.	Jack Plane Handles, bored for Bolts, varnished, for Iron Planes,	.65
3.	Fore or Jointer Plane Handles, for Wood Planes, ...	.75
4.	Fore or Jointer Plane Handles, with Bolts, for Wood Planes, . .	1.50

SAW HANDLES.

MADE FROM SELECTED STOCK.

Packed in Paper Boxes, One Dozen in a Box.

No. 5.

5.	Full size, Beech, Plain Edges, Slit only, ...	1.20
6.	Full size, Beech, Varnished Edges, Slit only, ...	1.30
6½.	Large size, Beech, Var. Edges, Slit only, for 30 to 32-inch Saws,	1.60
7.	Full size, Beech, Varnished Edges, with Brass Screws and Slit,	6.00
8.	Full size, Cherry, Varnished Edges, Slit only, ...	1.65

SAW HANDLES.

MADE FROM SELECTED STOCK.

Packed in Paper Boxes, One Dozen in a Box.

No. 9.   No. 10.

No.   Per Dozen

9.   Small Panel, Beech, Varnished Edges, for 16 to 20-inch Saws,

Slit only, ... $1.30

10.   Back Saw, Beech, Varnished Edges, Slit only, ...   1.20

No. 11.   No. 12.

11.   Butcher Saw, Beech, Varnished Sides and Edges, Slotted, . .   1.65

12.   Butcher Saw, Beech, Varnished Edges, not Bored or Slit, . .   1.25

No. 13.   No. 14.

13.   Compass Saw, Beech, Varnished Edges, Bored and Slit, . . .   1.10

14.   Compass Saw, Beech, Varnished Edges, Slit only, ...   1.00

SAW HANDLES.

MADE FROM SELECTED STOCK.

Packed in Paper Boxes, One Dozen in a Box.

No. 15.

No.
15.   Nest Saw, Beech, Varnished Edges, Bored and Slit, ... Per Dozen $1.15

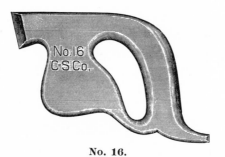

No. 16.

16.   One Man Cross-cut, Beech, Varnished Edges, Slit only, . . . 1.80

Prices will be quoted on Special Saw Handles to sample, on application.

WALL SCRAPER HANDLES.

In Paper Boxes, Three Dozen in a Box.

No. 17.

17.   Wall Scraper Handles, not Slit or Drilled, ...40

IMPROVED IRON SPOKE SHAVES.

N° 51

No.
51. Double Iron, Raised Handle, 10-inch, 2⅛-inch Cutter, . . . $3.50

N° 52

52. Double Iron, Straight Handle, 10-inch, 2⅛-inch Cutter, . . . 3.50

N° 53

53. Double Iron, Raised Handle, Adjustable Mouth, 10-inch, 2⅛-inch
Cutter, ... 4.50

N° 54

54. Double Iron, Straight Handle, Adjustable Mouth, 10-inch,
2⅛-inch Cutter, ... 4.50

N° 55

55. Double Iron, Raised Handle, Hollow Face, 10-inch, 2⅛-inch
Cutter, ... 3.00

N° 60

60. Double Irons, Double Cutter, Hollow and Straight, 10-inch,
1½-inch Cutters, ... 4.50

BOX SCRAPERS.

70. Box Scrapers, Malleable Iron, 2-inch Cutters, ... 6.00

PLANE IRONS.

MADE FROM SUPERIOR REFINED CAST STEEL.

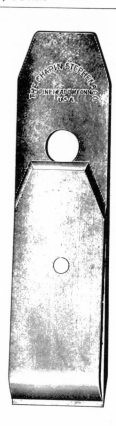

Single Cast Steel Plane Irons or Cut Irons.		Double Cast Steel Plane Irons.	
Width	Per Dozen	Width	Per Dozen
1½ inch, . . .	$2.50	1½ inch, . . .	$5.50
1⅝ " . . .	2.75	1⅝ " . . .	6.00
1¾ " . . .	2.75	1¾ " . . .	6.00
1⅞ " . . .	2.95	1⅞ " . . .	6.25
2 " . . .	3.15	2 " . . .	6.45
2⅛ " . . .	3.35	2⅛ " . . .	6.65
2¼ " . . .	3.70	2¼ " . . .	7.00
2⅜ " . . .	4.10	2⅜ " . . .	7.40
2½ " . . .	4.50	2½ " . . .	8.20
2⅝ " . . .	4.90	2⅝ " . . .	9.00
2¾ " . . .	5.50	2¾ " . . .	10.15
3 " . . .	7.00	3 " . . .	11.70
Assorted, 2 to 2½ inch, . .	3.75	Assorted, 2 to 2½ inch, . .	7.15

CAST STEEL GROOVING PLOW BITS.

Per Set, eight Irons, ⅛ to ⅝-inch, ... $1.50

Moulding, Rabbet, and Match Irons and Bits, all sizes in stock.

PLANE STOPS.

	Per Dozen
Plow Screw Stop, ...	$5.00
Dado Screw Stop, ...	4.00
Filletster Screw Stop, ...	7.00
Plow Side Stop, ...	1.80
Dado Side Stop, ...	1.80
Filletster Side Stop, ...	3.00